Trust Management in Virtual Work Environments

A Human Factors Perspective

Ergonomics Design and Management Theory and Applications

Series Editor
Waldemar Karwowski
Industrial Engineering and Management Systems
University of Central Florida (UCF) – Orlando, Florida

Trust Management in Virtual Work Environments: A Human Factors Perspective
Wiesław M. Grudzewski, Irena K. Hejduk, Anna Sankowska, and Monika Wańtuchowicz

Forthcoming Titles

Ergonomics and Psychology: Developments in Theory and Practice
Olexiy Ya Chebykin, Gregory Z. Bedny, and Waldemar Karwowski

Ergonomics in Developing Regions: Needs and Applications
Patricia Anne Scott

Handbook of Human Factors in Consumer Product Design, 3 vol. set
Neville A. Stanton and Waldemar Karwowski

- Human Factors Interaction Theories in Consumer Product Design
- Human Factors Methods, Guidelines, Standards, and Principles in Consumer Product Design
- Human Factors Design Case Studies in Consumer Product Design

Nanotechnology, Occupational and Environmental Health and Safety
Ash M. Genaidy and Waldemar Karwowski

Trust Management in Virtual Work Environments

A Human Factors Perspective

Wiesław M. Grudzewski
Irena K. Hejduk
Anna Sankowska
Monika Wańtuchowicz

CRC Press
Taylor & Francis Group
Boca Raton London New York

CRC Press is an imprint of the
Taylor & Francis Group, an **informa** business

CRC Press
Taylor & Francis Group
6000 Broken Sound Parkway NW, Suite 300
Boca Raton, FL 33487-2742

First issued in paperback 2019

ISBN-13: 978-1-4200-6890-0 (hbk)
ISBN-13: 978-0-367-38735-8 (pbk)

Library of Congress Cataloging-in-Publication Data

Trust management in virtual work environments: a human factors perspective /
Wieslaw M. Grudzewski ... [et al.].
p. cm. -- (Ergonomics design and management : theory and applications ; v 2)
Includes bibliographical references and index.
ISBN 978-1-4200-6890-0 (alk. paper)
1. Information technology--Management. 2. Business ethics. I. Grudzewski, Wieslaw.

HD30.2.T79 2008
174'.4--dc22 2008004770

Visit the Taylor & Francis Web site at
http://www.taylorandfrancis.com

and the CRC Press Web site at
http://www.crcpress.com

This book is dedicated to the memory of
the great management guru —

Peter F. Drucker (1909-2005)

"...we may safely trust a good deal more than we do. We may waive just so much care of ourselves as we honestly bestow elsewhere."

Henry David Thoreau

Walden: Or Life in the Woods, 1854

Contents

Contents

List of Figures

List of Graphs

List of Tables

Foreword

Human factors discipline focuses on optimizing the interactions between people and technology by adapting the human-centered approach to system design and management. This paradigm also applies to contemporary business systems that apply information technology and especially to the virtual organizations, which rely upon complex interactions of trust between people, organizations, and technology. Indeed, the success or failure of many virtual businesses, which operate across geographical and traditional organizational boundaries, depends to a large extent on the quality of individual and organizational trust.

Since virtual organizations are built upon the information technology, the issue of trust between people interacting with such technology, as well as trust of people in the information technology became an important consideration among many organizational and human factors studies. However, such studies were mainly disjointed in their discussion and treatment of the role of trust in the information age. The current book, written by Professors Wieslaw Grudzewski and Irena Hejduk, and their talented graduate students and Fulbright Fellows, Anna Sankowska and Monika Wańtuchowicz, offers a unique and insightful perspective on the critical nature of the individual and organizational trust interactions taking place in the complex sociotechnical systems of the contemporary business world from the human factors perspective.

I would like to congratulate the authors for their efforts to bridge the gap between the organizational and human factors studies of the nature of trust, and its critical importance in the context of ethical and successful design, operation, and management of virtual organizations.

Waldemar Karwowski, Series Editor

IEMS, University of Central Florida
Orlando, Florida, USA
September 2007

Preface

"Business is built upon trust but could be ruined in a second"—Phonetrade Golden Rules (Phonetrade.com). This is a common truth. Hence, it seems to be the most natural need—the need for trust management in organizations. Virtual organizations (VOs) are especially exposed to difficulties and limitations in their activities if there is not enough trust among their partners, cooperants, and clients. For this reason, trust management is most developed in that kind of organization.

Today's market game is almost impossible to imagine without trust. Kenneth J. Arrow, the economist and Nobel Prize winner (1972), opines justifiably that:

> Trust and similar values, loyalty or truth-telling, are examples of what the economist would call "externalities." They are goods, they are commodities; they have real, practical, economic value; they increase the efficiency of the system, enable you to produce more goods or more of whatever values you hold in high esteem. But they are not commodities for which trade on the open market is technically possible or even meaningful.…It has been observed, for example, that among the properties of many societies whose economic development is backward is a lack of mutual trust. (Arrow, Kenneth J., *The Limits of Organization.* New York: W. W. Norton and Company, 1974).

The topicality and the importance of issues connected with trust management inspired the authors to write this novel publication, convinced that trust is about to be a hot topic in the business world. This book is the very first elaboration in this field, a cross-sectional answer for global current trends in management. At the same time, this is the first attempt to systematize current knowledge about the multidisciplinary phenomenon of trust. The authors' intention is to present the issues of trust in management sciences in an interesting and perspicuous way.

Wiesław M. Grudzewski
Irena K. Hejduk
Anna Sankowska
Monika Wańtuchowicz

About the Authors

Prof. Wiesław M. Grudzewski, PhD, Eng., is a professor at the Warsaw School of Economics. He was visiting professor at California State University at Fresno (1994) and Seattle Pacific University (2001), and a visiting scholar at Robert Morris University School of Business in the fall of 2007. He is a distinguished member of many international scientific and professional associations, as well as a correspondent member of the Polish Academy of Sciences. He is chief editor of the monthly journal, *Economics and Organization of Enterprise.* He has authored 40 books and published more than 400 articles and research papers in the field of knowledge management, technological entrepreneurship, and consulting for small- and medium-sized enterprises.

Prof. Irena Hejduk, PhD, DSc, is the head of the Management Systems' Department of the Warsaw School of Economics, and professor at the Warsaw University of Technology. She is a visiting professor at University of Colonia (1990-1991), Seattle Pacific University (2001) and Robert Morris University School of Business (2007), as well as a visiting scholar at Robert Morris University (fall semester 2007). She is a member of many international and Polish scientific associations, and has published 35 books as well as more than 300 articles and research papers in the field of knowledge management, technological entrepreneurship, and consulting for SMEs.

Anna Sankowska and Monika Wańtuchowicz are PhD candidates at the Warsaw University of Technology, and visiting Fulbright Junior Grant researchers at Pennsylvania State University, as well as the Visby Programme at Stockholm University. Members of many scientific associations, their main fields of interest are virtual organizations, trust in management sciences, and innovations. In addition to publishing a book about trust in VOs, they have written chapters in two other books, and authored more than 30 articles in Polish and foreign conference proceedings and journals.

1 Theoretical Framework for Trust Management

1.1 ETHICAL UNDERSTANDING OF TRUST AND MANAGEMENT ETHICS IN GREAT WORLD RELIGIONS

ETHICAL ASPECTS OF TRUST

Trust involves far more than simply relying on another party. What makes this true is that trust might originate from notably good will, encapsulated interest, or compelled by the force of norms. Trust occurs under a variety of conditions, relates to many spheres of individual and collective life, and embraces a broad range of activities (Podrez 2003, p. 87). Moreover, it may stem from reasonable activities addressed by certain groups or individuals as well as from the presence of people in our proximate environment (family, friends) or those farther away (colleagues, public people). Gaining someone's trust at the beginning of a relationship is easier than retaining it. As such, trust is central to our lives, partially because the special concern we have for it must have a place within a broader set of moral concerns. Further more, this phenomenon is indissolubly associated with what is good and noble in man in a genuine sense (Graff 2003, p. 102). This is a very apparent ethical facet of the trust phenomenon. For this reason, trust and ethics are coupled into one topic for discussion.

Two theoretical points of view on trust shall be elucidated here. The first concerns man as a source of trust and posits that every human being is open, ready to trust and accept another person. The second directly mirrors social and economical life, in which trust is recognized together with certain aims (Wiśniewski 1996, p. 100–101). Along with these two perspectives, we can discuss ethical and practical aspects of trust. Ethical components originate from the fact that when we decide to trust somebody or expose ourselves to risk, we give this person something very valuable. As reciprocation, we expect loyalty and honesty. However, we should not apprehend trust only in the categories of mutuality and claims because then we can obtain only one from many possible transactions. Trust is simultaneously a sign of respect and a specific demonstration of humanity. Conversely, *mistrust* means that we negate somebody's humanity (Graff 2001, p. 309–316). The second practical component of trust means that the motive for trust is "commercial" transaction (Hardin 1991, p. 185–209). This transaction aims to maximize profits, and trust should be helpful in accomplishing that goal. One could argue that, in such cases, people's behaviors are predictable and plausible, but not trustworthy in a strict sense. Ethics is inevitably connected with the phenomenon of trust.

Ethics (gr. *ethos* – habit, *ethikos* – moral) is a philosophical science dealing with various moral issues—for example, moral norms, patterns, behaviors, values, and assessments (Bittner and Stępień 2000, p. 5). Socrates initiated ethical discussions; however, the term *ethics* was introduced by Aristotle.* According to Pratley (1998), ethics has many goals, including:

- To assess human activities referring to moral standards
- To guide moral behavior in certain situations

Ethics can be divided into **individual ethics** (emphasizes the privacy of every human) and **social ethics** (describes the group). There are also different agendas for **general ethics** (normative ethics) and **detailed ethics** (for instance, economic ethics). Economic ethics explains the morality of managing society and gives some values (it has an estimative character). Business ethics, a subsection of economic ethics that includes market economy ethics and economic life ethics, is a more detailed discipline that considers ethics as it is applied in business. According to Gasparski (2002), business ethics is the part of philosophical reflection on the ethos of businessmen. It aims to solve problems connected with questions regarding how to act. Hence, it is practically a new, distinct discipline (Chryssides and Kaler 1999). According to objective criteria within economic ethics, we can distinguish the following types:

- **Macroethics**—investigates national economy and society.
- **Mezoethics**—deals with an analysis of smaller economic units and other market institutions.
- **Microethics**—refers to individuals' activities in the sphere of markets.
- **Professional ethics**—analyzes morality practiced by specific professionals or occupations connected directly or indirectly with market economy (for example, insurance agents, tax advisors, teachers, doctors).
- **Global ethics**—is concerned with international economic activity and issues originating in cultural diversity.
- **Management ethics**—assesses the behavior of corporations and their managers as well as the levels on which particular functions of management are realized (Pogonowska 2004).

Ethical problems in corporate management are the main factors impacting companies' future growth and existence (Kaibori and Pawlak 1999). These problems can occur in all types of corporations, acting in different places all over the world. Management ethics is especially relevant to virtual organizations (VOs), which are based on mutual trust between partners.

Management of all organizations, including VOs, should be conducted in an honest manner. If this were the case, establishing a code of business ethics would become unnecessary. But economic practice *de facto* shows something totally

* The terms *ethics* and *morality* are commonly seen as synonymous, which is wrong. *Morality* concerns acts and decisions, whereas *ethics* is a theoretical reflection on morality.

different. Trust plays a crucial role in any form of economic activity, among business partners as well as within society (cooperants, consumers, and clients). Therefore, there is a need to respect ethics principles in management. Business ethics as a separate discipline has existed for only a short period of time, several decades. Similarly, ethical codes have been used in organizations for only about 20 years.

The most natural source of morality is religion. God decides what is good or bad and is an unquestionable authority on ethical issues. The consequence of this fact is that business people build their ethics on religion (Chryssides and Kaler 1999). A comparison of ethics in four great world religions—Judaism, Buddhism, Islam, and Christianity—is a good beginning for deliberation in the field of management ethics. The differences in religion, historical and cultural background issues challenge for management ethics. Religions still play a significant role and have an enormous impact on economic activity. Furthermore, progressive globalization, ecological issues, and poverty have begotten new ideas on the aims of enterprise and management coupled with acceptable behaviors and attitudes within an organization. All mentioned religions support entrepreneurship and aspiration to gain a profit. At the same time, they establish more-or-less detailed conditions and limitations. Employees' rights should be respected. In addition, entrepreneurs should fight their "natural" greed and profit ought not be their only goal; instead, it should be a means to another goal. The overall main target is client satisfaction.

Ethics in Virtual Organizations

Ethics in the burgeoning information society is a factor accelerating progress and, at the same time, is a restraint for contemporary organizations, such as VOs.* The nature of VOs literally drives alterations in the interpretation of the ethical issues existing in organization management.

It is worth noting the following two key issues of ethical background in VOs:

1. Ethics is the element in long-term functioning that determines a VO's existence, survival, and future development in the global market.
2. Ethics plays an important part in VOs, even more important than in traditional organizations. (Kisielnicki and Ochinowski 2005).

Three ethical issues are pertinent to VOs:

1. One or more parts of a VO failing to fulfill the conditions of the agreement.
2. Annexing an organization that aims to compromise another partner in the VO (for instance, by degrading or debasing a trademark of its products or services) or represents the interests of the competitors
3. Acting against the interests of their home countries when the partners of a VO work together to avoid embargo or other legal limitations. (Kisielnicki, *Problemy etyczne w organizacjach wirtualnych*)

* More details about VOs can be found in chapter 3, entitled "*Virtual Organizations.*"

The cardinal weight of ethics in VOs is discussed by Kisielnicki and Ochinowski (2005). The results of their survey illustrate the significance and complexity of the ethical problems (Graph 1.1). Graph 1.1 reveals crucial issues of ethical behavior in VOs. The biggest fears arise from concerns about the quality of the information about the client who is the initiating party for a VO (43% respondents) and the profitability of VO operational activities (42%). The skeptical attitude toward a VO originates from the creditability of the VO and its members (37%). It is essential to keep in mind the fact that creditability is often recognized as the equivalent (synonym) of trust, whereas ethics in organizations refers to reliability in different areas. To the main ethical issues in VOs, we should also add legitimacy of the undertaking (28%) and the expectations of partners (20%). Fewer doubts come from issues related to initial investment input (15%) and reasons for the selection of the organization to take part in a VO (11%). Other problems were pointed out by less than 10% of respondents and encompass some very detailed conditions of VO functioning. In summary, the results of the above survey confirmed without any doubt that ethical issues are crucial in VOs.

Contemporary organizations, like VOs, function in very turbulent environments. As as a result, ethics has become an essential managerial tool (Micklethwait and Wooldridge 2000). As a consequence of ethical deliberations in the field of management sciences, the enactment of trust is given much attention in organizations. In VOs, ethical issues and trust seem to be the greatest problems because of the specific

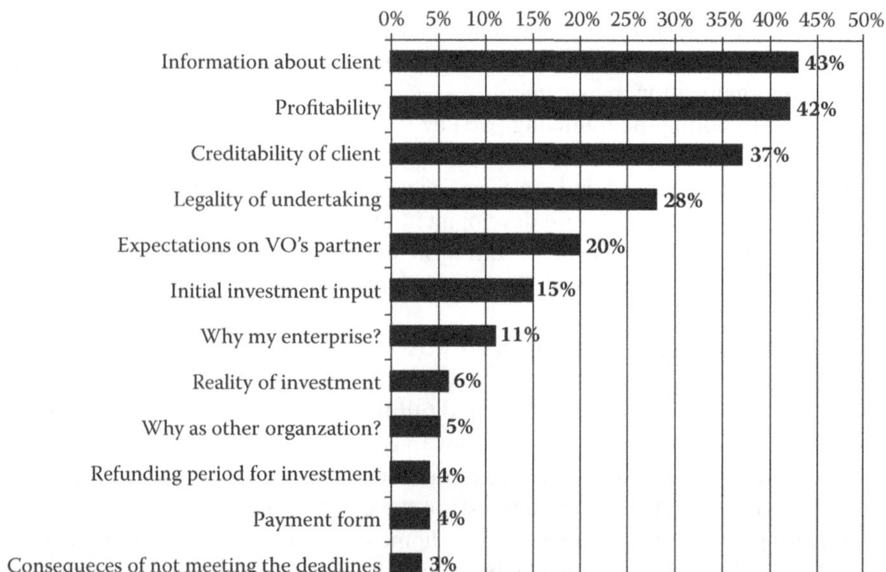

GRAPH 1.1 Crucial ethical problems in VOs. *Source:* Modified from Kisielnicki, J., and Ochinowski, T., 2005. *Etyka w organizacjach wirtualnych.* http://www.placet.pl/?mod=Artykulyandid=82andosCsid=23b3c48261d8fae06983f4f0f7c169b7 (accessed November 9, 2006).

character of VO activity.* Acting in the form of a VO provokes many temptations for unethical acts. The reasons for this are as follows:

- Inexperience with new partners (forming a VO)
- Fast pace of business activity
- Creating an "ad hoc" organization for the purpose of fulfilling only one certain assignment
- Lack of suitable legal regulations. (Kisielnicki, *Problemy etyczne w organizacjach wirtualnych*)

Additionally, the consequences of unethical activity may affect not only the unethical partner, but the whole VO. It is very hard to make VO partners responsible and accountable for something that happened. And this is the potential source of temptation.

Trust management is more important for VOs than for traditional organizations. It is not surprising then that VOs are leaders in trust management (Ramo 2004). Effective communication in VOs is one of the reasons for that.** Management of VOs differs significantly from management of traditional organizations. In traditional business, trust is usually not considered as relevant or as significant.

A range of problems related not only to ethics but also to trust in VOs is revealed in the very first stages of VO creation and operation. Future development of VOs depends mainly on the ability to overcome all obstacles linked with ethical behavior as well as on enactment of trust management. Therefore, it is easy to understand why this study begins with an analysis of ethical issues.

Ethics of Management in Great World Religions

Judaism

The fundamental book of moral principles for every Jew is the *Torah* (also called the *Law of Moses, Five Books of Moses,* or *Pentateuch*). *Torah* primarily refers to the first section of the Tanach (or Tanakh). The first five books of the Tanach are interpreted in the Talmud (a 36-volume record of rabbinic discussions pertaining to Jewish law, ethics, customs, and history). *"Leave your country, your family, and your relatives and go to the land that I will show you. I will bless you and make your descendants into a great nation."* (*Holy Bible: New International Version*, 1984, Genesis 12:1, 2). These words were addressed by Yahweh to Abraham. The very beginning of Judaism summons up patriarchs' times, which were approximately twenty-two centuries before Christ. Meaningful occurrences are migration through the desert, collection on Mount Sinai of the so-called Decalogue that is the Ten Commandments in the form of two stone tablets as well as the Sinaitic Covenant. "Decalogue is said to constitute the synthesis for the whole religion and morality of

---

* More about trust management in VOs can be found in chapter 4, entitled "Leading Examples of Virtual Organizations."

** More details about communication and its importance in trust management can be found in chapter 2, "Main Assumptions and Principles in Building Trust Systems," in the section entitled "Influence of Communication on Trust Building."

the Old Testament" (*Britannica*). Choice of Israel, Covenant choice and then law—all these occurrences led to announcing the Israeli to be the Yahweh (chosen) people (Jankowski 1985, p. 16). Covenant Law is the source of guidelines on how to run a business. Good fortune and failure in gaining material goods depends on God, which is stressed in everyday prayer (Kietliński, Reyes, and Oleksyn 2005, p. 19).

In agreement with the social justice rule, the rich help the poor directly or indirectly via charity institutions, which should be given at least the minimum level of living (de Lange 1996). Avoiding payment of taxes is treated as stealing from the poor. Moreover, it is unacceptable to squander any material goods that could be used by other members of society. Jewish ethics also forbids running a business that could be harmful to society or destructive to the environment. Monopolies and cartels are recognized as harmful to society (Minus 1998, p. 106).

Judaism rates very highly the work of human hands, which guarantees success, whereas laziness is viewed as the main cause of poverty. Living in penury is regarded as justice from God for human sins. That is why the poor were not respected in Israeli society.

It is worth taking into careful consideration the duality of some behaviors of Jews. For example, they are not allowed to borrow with profit from or cheat other Jews. However, this principle does not apply to people from other religions. According to the Talmud, it is also forbidden to deceive others in trade, accept bribes, and earn more than one-sixth part of a purchasing price (*Tarjag micwot, 613 przykazań judaizmu* 2000; *Z mądrości Talmudu* 1988).

Buddhism

The historical founder of Buddhism is said to be prince Siddhartha Gautama (first part of fourth century B.C.), who is also recognized by Buddhists as the Supreme Buddha of our age ("The awakened one") (Kotański 1986, p. 494). Siddhartha Gautama was a spiritual teacher and the best example to follow. He expounded the essence of Buddhist religious doctrine. But originally, Buddhism was seen as a way of life and practical philosophy, not as a religion. When there were effectuated the constructs of heaven and "the highest happiness," rituals were introduced too. The foundations of Buddhism are included and explained in Four Noble Truths. These four truths are:

- That existence is suffering (*dukkha*)—for instance, birth is suffering, aging is suffering, illness is suffering, death is suffering, union with what is displeasing is suffering, separation from what is pleasing is suffering, not to get what one wants is suffering
- That this suffering always has a cause and effect (*samudaya*)
- That it can be suppressed (*nirodha*)
- That there is a way (*magga*) to accomplish this. (Pratley 1998)

The principal moral characteristics in Buddhism are honesty, unselfishness, universal friendliness, and common compassion. The sources of suffering are thirsts and lusts. Buddhism aims to gain the state of perfection and proficiency called *nirvana* by means of the extinguishment of all desires that are the origin of suffering.

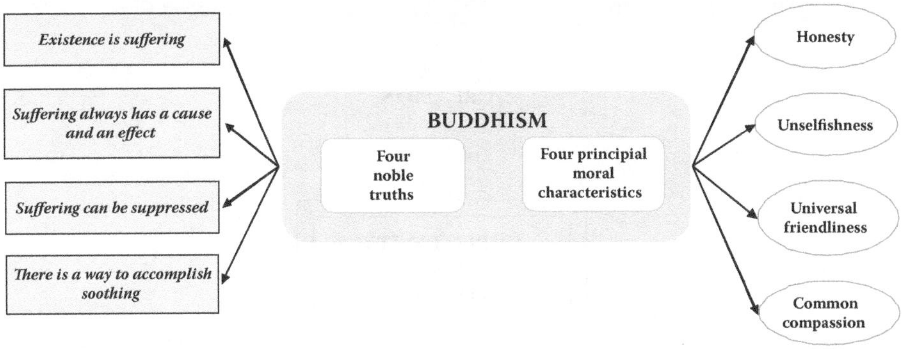

FIGURE 1.1 The underpinnings of Buddhism.

In reference to economic activity, the main principle is honesty. To achieve the state of nirvana, a human being must faultlessly do his job while striving for perfection in accordance with the following words: "Work diligently for your salvation" (Panikkar 1986, p. 116). When deciding on technology, one ought to take into account the well-being of employees; namely, one should use mainly labor-consuming technologies. It is advisable to produce goods that fulfill the needs of the highest number of people. Hence, production of goods that are luxurious is against the principles of Buddhism. Particular emphasis is put on cooperation of a common solution for important problems, such as poverty (Kietliński, Reyes, and Oleksyn 2005).

Buddhism is skeptical in assessment of economic growth in the term of value per capita. Average value of income might be misleading. Concurrent existence of wealth and poverty in one society is unacceptable. The priority is the general public well-being of all members of society. The process of economizing is crucial for economic growth. After fulfilling basic needs, one should restrict consumption and start economizing. At the same time, Buddhism forbids the investment of money in "black and gray" areas of economy that are sources of fast profits, for example, selling weapons, drug trafficking, and prostitution (Chryssides and Kaler 1999, p. 562).

Buddhism also defines the attitude of every human being to nature. We ought to respect the environment. We must not kill any animals if it is possible to obtain food by another way.

Japan illustrates a special case of economic success gained in accordance to the principles of Buddhism. Contrary to the highly developed economies of the European Union and the North America countries, Japan prioritizes serving to the whole society and protecting the environment. Geographical location, available natural resources, and enormous population have had a strong impact on the development of this collective way of thinking in Japan.

Christianity

The set of moral principles for all Christians is the Bible, a commonality with Jews. The very beginning of creation is associated with the appointment of Abraham as the

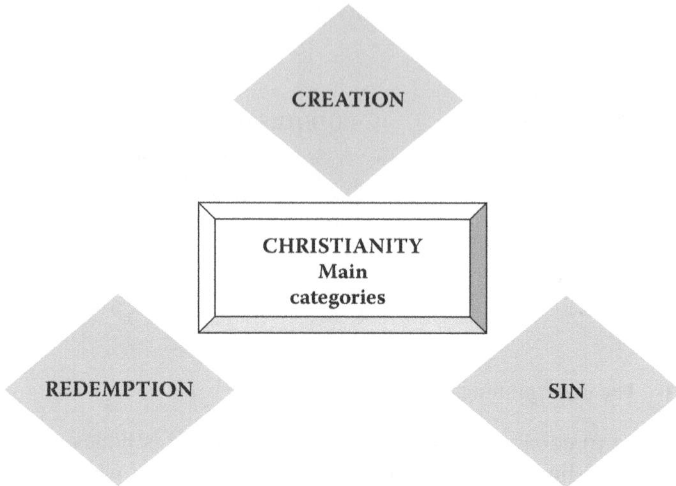

FIGURE 1.2 Main categories in Christianity.

Father of the Chosen Nation. New Moses became Jesus Christ. The quintessence of Christianity is built on three categories: creation, sin, and redemption (Figure 1.2).

It is assumed that business people should be submissive in all their actions (Häring 1966, p. 160). Contemporary capitalism sees every human being only as a consumer, a client, or a producer. Christianity supports a free market but under the condition that its main goal is to gain common welfare (Kietliński, Reyes, and Oleksyn 2005). In creating the concept of social market economy, Protestant and Catholic ethicists participated. In Genesis, God summoned the human being to work. Pope John Paul II highlighted the value of issues such as equitable pay, the right to create trade unions, and the right to have property. His encyclical *Laborem exercens* states that work is the focal dimension in man's life and the reason for his presence here on Earth. That kind of work leads to the development of humankind, and not just the multiplication of wealth (John Paul II 1996).

According to Christian doctrine, a man is not an owner but only a user of all material goods. Therefore, we need to follow the rule of social justice and the rule of man and brother's love. This means that "a well-being on the Earth" ought to be available for all people and not only for chosen individuals (Herr 1999).

John Paul II started a totally innovative look at profits of enterprise. High profit is seen as a sign of good functioning of an enterprise. It stems from suitable use of resources and fulfillment of human needs. Simultaneously, moral factors are essential for proper functioning of an enterprise. In light of transformations that are still taking place in Central-East Europe, entrepreneurship is a highly valued characteristic, equally with diligence, creditability, and reliability (John Paul II 1998).

Enormous input into contemporary ethics belongs to Martin Luther, a church reformer who founded Protestantism. According to Protestant thought, only people who are religious and ethical can run businesses. Hard work, scrupulous and honest doing duties of craftsman, merchant, manufacturer, banker, etc., restraining of expenses for pleasures and saving—these are the responsibilities of Protestantism

(Pogonowska 2004). Usury (taking a percentage from borrowed money) was considered to be intolerable. John Calvin was the main character who contributed to the development of Protestant ethics. In his opinion, serving God indissolubly embraced active economic activities. Wealth is a visible sign of God's grace, whereas poverty is the first sign of damnation (Stankiewicz 1998, p. 99). Moreover, Calvin accepted money making by clergymen.

Because of the situations of the countries belonging to it (superiority of agriculture), the Orthodox Church did not have too much impact on business ethics. "Blessed are you who are poor, for yours is the kingdom of God" (*Holy Bible: New International Version*, 1984, Luke 6:20). Condemnation of wealth originates from associating the rich with greed. In order to be God-fearing and fervent, it is necessary to reject all properties (*Katechizm Cerkwi prawosławnej* 2001). The Orthodox Church greatly appreciated the importance of farmer workers. At the same time, people who worked in factories and developed cities were strongly criticized. Usury was unacceptable but, in contrast to other Christian churches, there were no further bans from canonical law.

Catholicism, Protestantism, and the Orthodox Church held similar attitudes with regard to two issues: the dangers that result from excessive consumption and environmental contamination (Bułgakow 1992).

Islam

Islam, the youngest of the greatest monotheistic religions, has its origins in the person of Abraham. The major book for every Muslim is the Koran. It consists of 114 *surahs* (chapters) with a total of 6,236 *ayat* (verses). It is believed to be the literal word of God (Arabic Allah) as revealed to Muhammad. It contains the truth of faith, whereas the content of law (*szaria*) determines the moral rules and principles that should be respected in everyday life within the Muslim society. The last prophet was Muhammad, who is believed to be the founder of Islam. His teaching is said to be the act of transfer of eternal truth, which was shown to him by the only God Allah with help of the archangel Gabriel. Similar to Judaism, Christianity, and Buddhism, Islam admonishes people for doing anything to others that would be unpleasant for us (Bassiouni 1998).

The priority is given to public interest. All goods come from Allah and owning them is acceptable under the condition that they were gained in an honest way. There is a very old tradition that rich Muslims assign 2.5% of their fortune to help the poor (alms) (Kościelniak 2002). The alms are one of the five pillars of Islam (Figure 1.3); the others are as follows: confession of faith, prayer, fast, and pilgrimage to Mecca (Pogonowska 2004). It is forbidden to squander and abuse one's fortune.

For a long time, Arabs were said to be the nation of merchants. Even Muhammad was a merchant and the great example for every Muslim doing the same profession. Work is believed to be a very important part of human activities. The state should guarantee work for everyone who is willing to do it. Women and men ought to be given work and adequate pay. Profit is the sign of individual self-development and, at the same time, the development of the whole society (Kietliński, Reyes, and Oleksyn 2005). The Koran advises to gain balance between work and rest. It is necessary to reach the highest position within an enterprise, be punctual, be precise, and be open

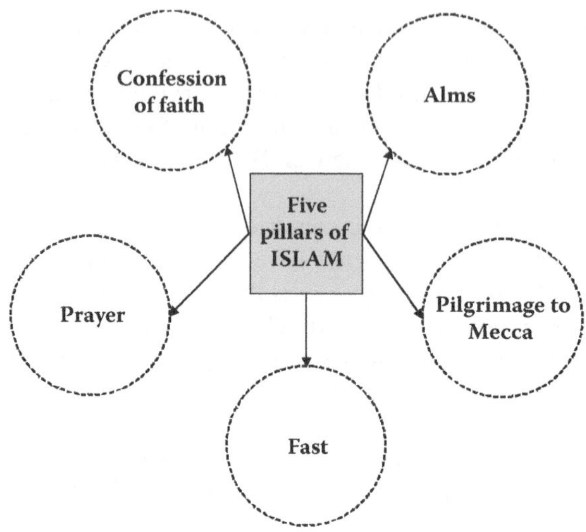

FIGURE 1.3 Five pillars of Islam.

to cooperation. Nowadays, the last feature is especially highlighted. The target of that kind of cooperation is to gain a common well-being, wealth. It is unacceptable for the sources of income to be gambling, growing rich on somebody else's harm, or using the complex situation of the poor. Usury is treated in Islam as profit gained without any work, and it is strongly intolerable (Klöcker, Tworuschka, and Tworuschka 2002, p. 18). But there is also one contentious issue: usury is not banned in contacts with representatives of other religions. Islam sanctions investment in capital in shares, trust funds, investment funds, and so on. "Spoken words" are treated in the same way as an agreement written on paper.

There are also other ethical concepts built upon religions other than the four main world religions, such as Hinduism, Confucianism, and Taoism.

REFERENCES

Bassiouni, M. C. 1998. Etyka w biznesie a islam. In *Etyka w biznesie*, P.M. Minus (ed.), Warszawa: PWN.

Bittner, B., and Stępień, J. 2000. *Wprowadzenie do etyki zawodowej*. Poznań: eMPi².

Bułgakow, S. 1992. *Prawosławie. Zarys nauki Kościoła prawosławnego.* Warszawa: Orthdruk.

Chryssides, G. D., and Kaler, J. H. (eds.). 1999. *Wprowadzenie do etyki biznesu.* Warszawa: PWN.

Gasparski, W. 2002. *Etyka biznesu – szkice do portretu.* In *Etyka Biznesu*, J. Dietl and W. Gasparski (eds.), Warszawa: PWN.

Graff, G. 2001. Czy istnieje cnota zaufania? *Przegląd Powszechny*, 12, 303–316.

Graff, G. 2003. Etyczne aspekty zaufania. *Prakseologia,* 143, 99–108.

Hardin, R. 1991. Trusting persons, trusting institutions. In *Strategy of Choice*, R. J. Zeckhauser (ed.), 185–209. Cambridge, MA: MIT Press.

Häring, B. 1966. *Nauka Chrystusa.* vol. IV, Poznań: Pallottinum.

Herr, T. 1999. *Wprowadzenie do katolickiej nauki społecznej.* Kraków: WAM.

Holy Bible: New international version. 1984. International Bible Society, Colorado Springs, CO.

Jankowski, A. 1985. *Biblijna teologia przymierza.* Katowice: Księgarnia Świętego Jacka.

John Paul II. 1981. *Laborem Exercens*. Available at www.vatican.va.
John Paul II. 1991. *Centesimus Annus*. Available at www.vatican.va.
John Paul II. 1996. *Encykliki ojca swiętego Jana Pawła II*. Kraków: Znak.
John Paul II. 1998. *Centesimus annus: Tekst i komentarze*. Lublin: Redakcja Wydawnictw KUL.
Kaibori, W., and Pawlak, W. R. 1999. O etyce w zarządzaniu przedsiębiorstwami w Polsce, *Problemy jakości*, no. 1/99.
Katański, W. 1986. Buddyzm jako religia światowa. In *Zarys dziejów religii*, J. Kelner et al. (eds.), Warszawa: Iskry.
Katechizm cerkwi prawosławnej. 2001. Hajnówka: Bratczyk.
Kietliński, K., Reyes, V. M., and Oleksyn, T. 2005. *Etyka w biznesie i zarządzaniu*. Kraków: Oficyna Ekonomiczna.
Kisielnicki, J. *Problemy etyczne w organizacjach wirtualnych*. http://www2.wz.uw.edu.pl/ksiz/download/Wirtualna-Etyka2005.pdf (accessed July 2, 2007).
Kisielnicki, J., and Ochinowski, T. 2005. *Etyka w organizacjach wirtualnych*. http://www.placet.pl/?mod=Artykulyandid=82andosCsid=23b3c48261d8fae06983f4f0f7c169b7 (accessed November 9, 2006).
Klöcker, M., Tworuschka, M., and Tworuschka, U. 2002. *Etyka wielkich religii: mały słownik*. Warszawa: Verbinum.
Kościelniak, K. 2002. *Chrześcijaństwo w spotkaniu z religiami świata*. Kraków: M.
Micklethwait, J., and Wooldridge, A. 2000. *Szamani zarządzania*. Poznań: Zysk i S-ka.
Minus, P. M. (ed.). 1998. *Etyka w biznesie*. Warszawa: PWN.
Panikkar, R. 1986. *Religie świata w dialogu*. Warszawa: PAX.
Podrez, E. 2003. Aksjologiczne i normatywne aspekty zaufania. *Prakseologia*, no. 143.
Pogonowska, B. 2004. Dziedziny refleksji etycznej nad gospodarowaniem. In *Elementy etyki gospodarki rynkowej*, B. Pogonowska (ed.). Warszawa: PWE.
Pratley, P. 1998. *Etyka w biznesie*. Warszawa: Gebethner and Ska.
Stankiewicz, W. 1998. *Historia myśli ekonomicznej*. Warszawa: PWE.
Tarjag micwot, 613 przykazań judaizmu. 2000. Kraków: Fundacja Ronalda S. Laudera w Krakowie.
Wiśniewski, R. 1996. Ontologia zaufania w biznesie. In *Etyczny wymiar przekształcen gospodarczych w Polsce*, A. Węgrzecki (ed.). Kraków: Cracovia.
Z mądrości Talmudu. 1988. Warszawa: PIW.

1.2 HISTORICAL OVERVIEW OF MANAGEMENT ETHICS DEVELOPMENT

Together with development of humankind, the need for establishing the ethical rules for economic activities is becoming more and more apparent. Changing philosophical trends, historical conditions, and many other factors influenced forming of the business culture. Currently, the driving force of civilization development is the economy rather than philosophy, religion, science, literature, or politics, which were significant before (Rybak 2002, p. 89). Management plays a significant role in building new systems of values for the contemporary human being.

Ancient Times

In ancient times, a discussion about law and medicine was started. A very good example of one of the very first codes is the so-called Hippocratic oath. All economic activities were associated with culture and religion. It was believed that human beings were created by gods to do work. Trade was linked with human greed but, simultaneously, it was understood that there is a need for exchange of goods. Traces of that kind of activity are known in the area of past Mesopotamia, where trade documents and law codes with laws from the first and second millennium B.C. belonging to Sumerians, Akkadians, Babylonians, and Assyrians were found (Kramer 1981). We should not forget about the famous code of Hammurabi from the seventh century B.C., which mainly regulated business activity, drawing up agreements, and all issues connected with owning property. The behavior of merchants of that time, which included using unfair methods to gain profits, determined the current lack of respect for this occupation. Legal procedures were the source of prosperity in business. Justice was at the top of the list.

Ancient Greece and Rome initiated the economical thought. During these times appeared many socioeconomic changes. The main activities were agriculture as well as industry, trade, and shipping. Then a shift occurred from economy based on goods (barter) to money economy. The interests of Greek ethicists of management were strongly connected with the issue of rarity of goods. There was a need for efficient production management, production maximization with minimization of costs (Stankiewicz 1998). The term *economics* appeared for the first time in Xenophon's *Oikonomikos* (meaning "household management"). It became a practical science containing the whole set of rules from the area of management. According to Xenophon, there is a need for state integration in order to gain a high rate of development. The most appreciated occupations were a farmer and a soldier. The keys to efficient management were proper division of labor as well as suitable usage of natural resources (Landreth and Colander 1998, p. 58).

According to Plato (427–347 B.C.), the priority is good. His concept of the ideal state divides the whole society into three layers:

- Rulers (philosophers)
- Guardians (soldiers)
- The rest of the people (farmers, merchants, slaves)

The two first classes of citizens were responsible for managing and making decisions—for example, establishing the pay limitations and prices of goods (Stankiewicz 1998). Merchants, craftsmen, and farmers belonged to the third class of hard-working people. Plato claimed that people making money are not able to gain any spiritual development. Simultaneously, he asserted that wealth helps gain the state of happiness, especially in older age. He suggested that inheritance is more noble that gaining any wealth as the result of hard work. Plato strongly criticized the practice of taking a percentage when lending money. According to his point of view, the state should regulate the amount of credits.

Aristotle is said to be the first economist (384–322 p.n.e.), and that is why many believe that he is also the first business ethicist. He separated two main fields in economics:

- The art of household management (*oeconomicus*)
- The ability to make money (*chrematiske*)

Household was the key element of family and the whole of society. Making money was done by so-called "social spongers" who lacked any virtue (Solomon 1997, p. 333, 402).

This activity, in the opinion of the very first economist, was totally worthless. According to Aristotle, both human needs (natural) and human cravings (impossible to fulfill) existed. He supported reaching the so-called "gold–middle." Work for money was the domain of citizens from the lowest social class. Every citizen had a right to freedom. In special cases—for example, when one was getting richer and richer on the poverty of others—he believed the state should interfere. Aristotle's opinion of merchants and tradesmen was sustained through many centuries and it was the main basis for ill-disposed opinions toward these occupations. Trade was often associated with usury (Solomon 1997, p. 333, 402). Experiences gained by every man taught him to be considerate and reasonable, which helped him to act in an ethical and trustful way (Aristotle 2002). Life teaches us what is of lasting value and whom we cannot trust.

Romans were eulogists in the field of agriculture, and this was reflected in the works of Cato and Cicero. All merchants, usurers (money-lenders), and tradesmen were treated as the lowest class of citizens by Romans and Greeks. Cato *au fond* considered usury to be much worse than stealing. Economy in the Roman Empire was regulated by detailed law. Ethical rules had priority over moral rules. To summarize the issues from ancient ethics, we can use the following words: "How without touching the virtue of wealth one may gain this wealth, how without resigning from this virtue one may make use from this wealth" (translated after Pogonowska 2004).

Christianity and Middle Ages

In the Middle Ages, man tended to care about religious issues, the necessity to make money, and the use of material goods. Riches were seen as something with very negative value. It was recommended to resign from any wealth and property. There were four main social classes: slaves, landowners, royal family, and church (Landreth and Colander 1998, p. 62).

Saint Thomas Aquinas was referring to the Gospel when he wrote: "So in everything, do to others what you would have them do to you" (*Holy Bible: New International Version* 1984, Matt. 7:12). Selling anything for a bigger sum of money than its real value was strictly forbidden (St. Thomas Aquinas 1981). The source of all the guidelines was the Bible. According to the Bible, it is impossible to serve God and money at the same time: "No one can serve two masters. Either he will hate the one and love the other, or he will be devoted to the one and despise the other. You cannot serve both God and money." (*Holy Bible: New International Version* 1984, Matthew 6:24).

The essence of trade is to buy at a low price and to sell at a higher price. This meant that trade was always connected with sin. But the New Testament pronounces that gaining any benefit from money is evil (*Holy Bible: New International Version* 1984, Matthew 25: 26–27). For usury in the Middle Ages, there was retribution. Usurers could not be granted absolution and also could not take communion. Furthermore, they were not allowed to obtain a Church funeral (Kietliński, Reyes, and Oleksyn 2005, p. 84). As a consequence, this occupation was the domain of Jews who could not do any other work. That is the way that trade and banking became the domains of Jews (de Lange 1996).

Another change introduced in the Middle Ages was supporting trade on an international scale. There were trials to set a fair price of money. The following words of Montesquieu had great significance: "Only competition enables us to determine the fair price of goods and establishes a fair relationship between them. Money needs to be reasonably priced. It is too high if tradesman do not attempt to purchase it." (Montesquieu 1748/2001).

Modern History and Contemporaneity

A. Smith or T. R. Malthus (classical school), J. Benthamow, J. S. Mill, J. B. Say (liberal school), J. M. Keynes, J. K. Galbraith, and M. Friedman were among the famous economists who made it possible to establish economics as a separate field. Existing philosophical concepts were interpreted once more in a new way. It is said that capitalistic ethics were created by Protestantism (Weber 1925/1999). Wealth was seen as something useful and not as a sign of a materialistic attitude (Margolis 2003). Currently, we expect enterprises and their managers to secure our job positions and the general common well-being. When usury was removed, then credit appeared. The price of goods and services was determined by demand and supply on the free market.

Kant (1724–1804) opined that the entrepreneur should not borrow money if he was not able to repay it. His ethical system was based on a principle called *categorical imperative* which means: Act only according to that maxim by which you can and hope that it would become a universal law (translated after Pogonowska 2004, p. 72 and next). General norms should be the same for all people and all enterprises without exception. Moreover, we need to respect each human being and not use him as a means to meet a certain target: "Act in such a way that you always treat humanity, whether in your own person or in the person of any other, never simply as a means, but always at the same time as an end" (Kant 1785/2001; Jackson 1999). Kant also strongly opposed monopolies in economy because they create new conflicts of an ethical nature. For instance, if a certain monopoly possesses a patent for a life-saving drug, it should never cease producing it. But this denies monopolies the opportunity to make choices (Klimczak 2002, p. 54).

A totally different standpoint was proposed by utilitarianism and its secular ethical representatives. The beginning of this system was associated with the person of F. Hutcheson (1694–1746), the author of the following principle: The ultimate goal of all human effort is to secure happiness for as many people as possible (Pogonowska 2004). The founder of utilitarianism was J. Bentham (1784–1831), who was in favor of the so-called "rule of utility." He posited that pain and pleasure are the only two absolutes in the world. His rule of utility is that "the good is whatever brings the greatest happiness to the greatest number of people." Later, he dropped the second part and talked simply about the greatest happiness principle. Therefore, any act that can save enterprise is ethical. Man is naturally selfish. He tries to fulfill his own needs. Summing up, all the attempts of every individual are the source of social development. Man is even able to support the interests of the whole of society if it will help him advance his own interests (Smith 1776/1998). Economic ethics plays the second role. The first role is gaining profit by enterprise. It is not rare that these two aims are opposite to each other; there is a collision of aims. Ethics in utilitarianism is strongly connected with the concept of an economic human being (*homo oeconomicus*). According to this concept, the acts of all people result from the rule of utility, which means here maximizing one's own well-being and wealth.

Work from the foundations and organic work were the subjects of interest in the works of Polish positivism writers (such as Orzeszkowa, Prus, Świętochowski). These works highlighted the need to save, sense of duty, judiciousness, entrepreneurship, and so on. These characteristics were treated as moral virtues during positivism (Kietliński, Reyes, and Oleksyn 2005).

A strong supporter of a very extreme view on economic ethics was Milton Friedman. In his opinion, there is no need for any ideology. It is enough to act on the free market without any traps and deceptions in order to simply act ethically. In business, it is enough to have just "ordinary decency." Moreover, the main goal of every manager is to maximize profit: "The business of business is business" (translated after Friedman 1997, p. 49).

Establishing business ethics as a separate field is associated with the publication of the book by W. Everett, *The Fundamentals of Business Ethics,* in 1926. It is the youngest ethical discipline. Its development was continued through the sixties of the last century and the period of the greatest development was in the eighties (Dylus 2000). The milestones for business ethics were various occurrences—for example, founding the *Transparency International* world organization to fight corruption, building the first code of ethics during the conference in Caux in 1994, and giving the Nobel Prize to Amartya Kumar in 1998.

Currently, more and more often we pay greater attention to ethics of management than to business ethics. Management ethics is replacing business ethics. It is caused by continued development of management sciences as well as the practical background of management. **Management ethics** deals with ethical norms that influence the behaviors of a whole organization and are the final results of actions of many members from the organization (Kisielnicki and Ochinowski 2005; see also Griffin 1995). In order to measure the value and importance of management ethics to the functioning of contemporary organizations, we ought to look at the damage caused by its lack or too low level (Sternberg 2000). Garrett and Klonoski (1986) distinguished three main areas within management ethics: attitude of organization to employee, attitude of employee to organization, and attitude of organization to other businesses, especially to clients and collaborating organizations. The influence of each organization member on organization behavior might be very various and diverse.

REFERENCES

Aristotle. 2002. *Nicomachean ethics.* Newbury, MA: Focus Publishing/R. Pullins.
de Lange, N. 1996. *Wielkie kultury świata. Żydzi.* Warszawa: Świat Książki.
Dylus, A. 2000. MoŻliwości i granice kazuistycznej etyki biznesu. In *Etyka biznesu,* J. Dietl and W. Gasparski (eds.). Warszawa: PWN.

Friedman, M. 1997. Społeczną powinnością biznesu jest pomnaŻanie zysków. In *Etyka biznesu: z klasyki współczesnej myśli amerykańskiej*, L.V. Ryan and J. Sójka (eds.). Poznań: W drodze.
Garret, T. M., and Klonoski, R. J. 1986. *Business Ethics*. Englewood Cliffs, NJ: Prentice-Hall.
Griffin, R. W. 1996. *Podstawy zarządzania organizacjami*. Warszawa: PWN.
Holy Bible: New international version. 1984. International Bible Society, Colorado Springs, CO.
Jackson, J. C. 1999. *Biznes i moralność*. Warszawa: PWN.
Kant, I. 1785/2001. *Groundwork of the metaphysics of morale*. (*Grundlegung zur metaphysik der sitten*). New York: Cambridge University Press.
Kietliński, K., Reyes, V. M., and Oleksyn, T. 2005. *Etyka w biznesie i zarządzaniu*. Kraków: Oficyna Ekonomiczna.
Kisielnicki, J., and Ochinowski, T. 2005. *Etyka w organizacjach wirtualnych*. http://www.placet.pl/?mod=Artykulyandid=82andosCsid=23b3c48261d8fae06983f4f0f7c169b7 (accessed November 9, 2006).
Klimczak, B. 2002. *Etyka gospodarcza*. Wroclaw: Wyd. AE im. Oskara Langego we Wrocławiu.
Kramer, S. N. 1981. *History begins at Sumer: Thirty-nine "firsts" in recorded history*. Philadelphia: University of Pennsylvania Press.
Landreth, H., and Colander, D. C. 1998. *Historia myśli ekonomicznej*. Warszawa: PWN.
Margolis, J. 2003. Chytry dwa razy zyskuje. *Forum*, 52, 28.
Montesquieu, C.. 1748. De l'esprit des lois (The spirit of the laws), Montesquieu, Charles de Secondat. 2001. *Montesquieu's science of politics: Essays on the spirit of laws*. Lanham, MD: Rowman and Littlefield Publishers.
Pogonowska, B. 2004. Dziedziny refleksji etycznej nad gospodarowaniem. In *Elementy etyki gospodarki rynkowej*, ed. B. Pogonowska. Warszawa: PWE.
Rybak, M. (2002). Społeczna odpowiedzialność przedsiębiorstw. In *Etyka w biznesie*, M. Borkowska, and J. W. Galkowski (eds.). Lublin: TNKUL.
Solomon, R. C. 1997. Arystotelesowskie podejście do etyki biznesu. In *Etyka biznesu: z klasyki współczesnej myśli amerykańskiej*, L.V. Ryan and J. Sójka (eds.). Poznań: W drodze.
Smith, A. 1776/1998. *An inquiry into the nature and causes of the wealth of nations*. Washington, D.C.: Regnery Pub.
Stankiewicz, W. 1998. *Historia myśli ekonomicznej*, Warszawa: PWE.
Sternberg, E. 2000. *Just business: Business ethics in action*, New York: Oxford University Press.
St. Thomas Aquinas. 1981. *Summa theologica*, Westminster, MD: Christian Classics.
Weber, M. 1925/1999. *Wirtschaft und gesellschaft* (*Economy and society*) Tübingen: J.C.B. Mohr (Paul Siebeck).

1.3 THEORETICAL APPROACHES TO TRUST

In order to understand dynamic interactions between people and social systems, scientists are looking for an answer in the social sciences to the following question that has bothered them since it was formulated by Harold Lasswell in 1958: "Who gets what, when, and how?" Since that time, in social sciences there has existed the conviction that power is the driving force that motivates us to take part in politics and also explains some practices and results gained by organizations.

Nonetheless, there appeared voices questioning the primacy of power answering the question, suggesting the fundamental role played by trust (Arrow 1997, Lipset and Schneider 1983, Gibb 1978, Luhmann 1979, Misztal 1996). Trust has become the subject of interest in the light of many phenomena: why some people, some organizations, and even some nations are more efficient and why in some nations virtues such as honesty, clearness, and faith in social development are more consolidated than in others (see, for example, Fukuyama 1995; Putnam, Leonardi, and Nanetti 1993; Whitney 1996; and Harisalo and Miettinen, 1997). Fukayama (1995) states that trust is creating social capital whose value determines the economic results for the whole society. Based on historical analysis, countries with a low role of trust were more likely to gain lower total economic gain than countries with a high role of trust. Knack and Keefer (1997) empirically proved the hypothesis of Fukuyama (1995) based on a trust index derived from the World Values Survey. They discovered a strong positive correlation between the appreciation of trust in a society and the society's economic growth. Trust determines the whole society and its economic results, as shown in Figure 1.4.

Generalized trust stands for trust between strangers and is not based on previous cooperation. Indices for generalized trust gained by Knack and Keefer (1997) were built on the question asked in 32 market economies: "Generally speaking, would you say that most people can be trusted or that you don't have to be very careful in dealing with people?" (Sztaudynger 2003, p. 233). Thirty-two percent of respondents answered that "they can trust most of the people" (the lowest percentage was in Peru [5.5 %], and the highest was in Norway [61.2%]). Based on these research results and using the method of ordinary least squares (OLS) regression on trust, Zak and Knack (as cited in Sztaudynger 2003, p. 235) prepared the following model of the income increase per capita (average for the years 1970–1992):

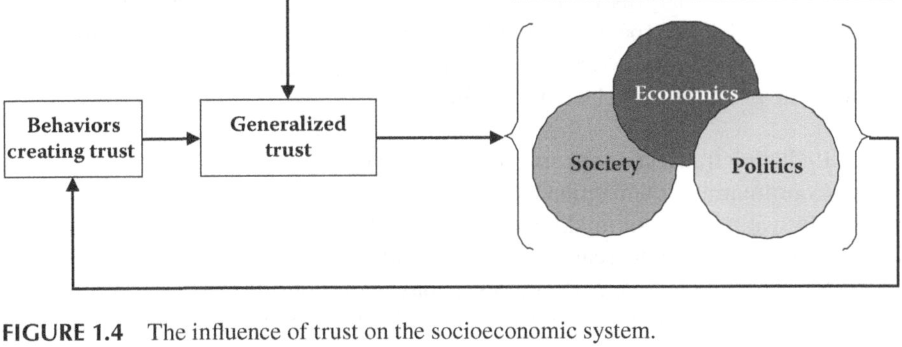

FIGURE 1.4 The influence of trust on the socioeconomic system.

$$X/C = 1{,}1 + 0{,}14\ I/X - 0{,}20\ X/C - 0{,}09\ SCHOOL - 0{,}024\ pI + 0{,}045\ TRUST \quad (1.3.1)$$
$$\quad (1{,}5)\ (4{,}4)\quad (2{,}0)\quad (1{,}0)\quad (3{,}0)\quad (2{,}6)$$

$R^2 = 57.0\%$ $S_e = 1{,}1$

where:

X/C	average rate of income growth *per capita* in the years 1970–1992
X/C	initial level of income *per capita* in 1970, thousands dollars
I/X	investment rate, average in the years 1970–1992
pI	process of investment goods in % in the United States
SCHOOL	average number years of education for population over 25 years old
TRUST	generalized trust, % of respondents indicating that they can trust most people, 1981, 1990–1991, 1995–1996.

(In parenthesis, absolute values are given for t-student statistics.)

According to the above model, an increase in trust of 10% would cause a rise of the income growth rate *per capita* of about 0.5%. For the growth of the average economic growth dynamics in surveyed countries, this translates to approximately one-fourth.

Generalized trust appears in this model as "TRUST," and it obtains different levels according to society. TRUST is based on moral foundations and agreement of every individual that the rest of the society accepts the same moral principles. Generalized trust is a general propensity of a certain society and it classifies the society as more or less skeptical or more or less trusting in terms of business relations. This general propensity is of course then completed by the specific context of certain transactions.

As a consequence of broad acceptance of trust, the role of trust, as the element connected with the cooperation in turbulent conditions and difficult to predict environment, was also acknowledged in the management sciences. Many authors (e.g., Kouzes and Posner, 1993, 1995, Harisalo and Miettinen 1995, Lane 1998, Sydow 1998, Ciancutti and Steding 2000) even state that trust sets the foundations for new types of social relations. Lane (1998) goes further by saying that the variety of exchange relations and the existence of new business forms would not be possible without interpersonal trust and trust among organizations. Well-grounded trust is beneficial (functional) for trustee and trustor. Giving somebody trust is followed by temporary resignation from the moral limitations for operations. People, organizations, and institutions that are given "the trust credit" are not constantly controlled and gain broader margins for their nonconformist, innovative, and unusual activities. At a social scale, specific examples of aggregated trust lead to rising mobilization, activity, and innovativeness (Sztompka 2005, p. 321). Trust is also recognized through the prism of other benefits, for instance:

- Reducing transactional costs (Handy 1995)
- Coordinating organization (Shapiro 1987)
- Motivating decision-making (McAllister 1995)
- Setting into action creative thinking (Szulczewski 2003)
- Encouraging to take part in transactions (Dasgupta 1998)
- Promoting open information exchange (Earley 1986)

- Supporting existence of markets (Akerlof 1970)
- Increasing company abilities (as a team) to survive the situations (Filek 2003)
- Building networks for cooperation and social coherence
- Helping build the civic culture and democracy.

In the opinion of Dasgupta (1988), trust is the central factor for all transactions. Furthermore, trust, via the functions that it fulfills, manifests itself as a capital-like financial capital, knowledge capital, educational capital, cultural capital, and so on.

Frequently, trust is included in the category of **social capital** (Coleman 1988, Fukuyama 1995). According to the World Bank, "social capital" refers to networks and relations that encourage trust and mutuality as well as define quality and the quantity of social interactions (World Bank 2000). Putnam (1995) acknowledges as social capital:

1. Bases and values (especially mutuality norms)
2. Networks (containing relations)—according to Lin, Cook, and Burt (2001) they concern social relations between individuals, employees, teams, organizations, societies, regions, and nations helping gain positive results
3. Trust (Figure 1.5).

The elements of social capital allow participants to work more efficiently in order to accomplish a common targeted activity with better results. Groups with members who have complete or high trust in each other always get much better results than comparable groups with low trust or without trust at all. Social capital is a resource with highly unique and specific value that helps and fosters achieving various targets. Moreover, it has the tendency to multiply itself. Trust owners have that possibility to have more and more trust.

Since the nineties, social capital has been introduced to econometric growth models (Sztaudynger 2003). It encourages innovation, education, and self-education and is as important as human and physical capitals.

Furthermore, we cannot omit the linkages between two essential capitals in contemporary organization: **financial capital** and **trust capital.** *Financial capital* is

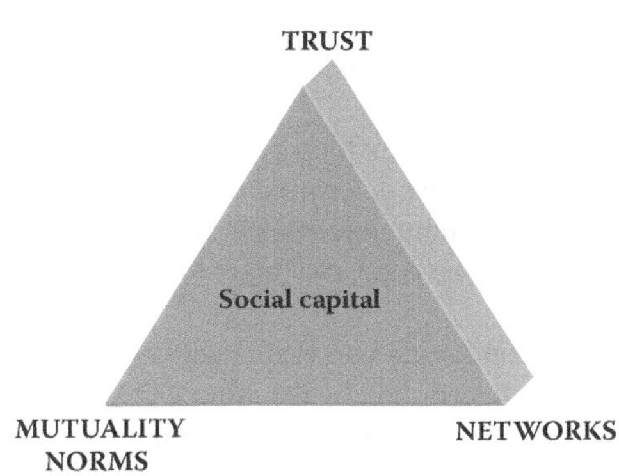

FIGURE 1.5 Social capital according to Putnam (1995).

usually defined as a critical factor without which production is impossible (Friedman 1986), whereas trust capital empowers and enables cooperation and decision-making, engenders the atmosphere of openness and transparency, enhances communication, and motivates and joins people together. Moreover, financial capital is available on the market as a product (e.g., credit) and has a certain value measured in monetary units. At the same time, trust capital is unique and is not a subject of trade. Establishing trust is time-consuming. These are some reasons why trust is said to be the source of a sustainable competitive advantage.

Sztompka (2007, p. 246–247) distinguishes two major constituents of trust capital that every individual can posses. The first one is **reliability capital,** which is received from others who trust the individual. The second one is **trustfulness capital,** which can be given to another person who an individual trusts. These two capitals complement and multiply one another. The more reliability capital we have, the more strangers will be induced to trust us. The more trustfulness capital we have, the more strangers will be trustful. For instance, the author who wrote the newest best seller gained trust from many readers and will more easily attract new readers. However, we cannot forget about the black side of social relations. At the same time, we can distinguish **distrust (mistrust) capital,** which is in opposition to trust capital (reliability and trustfulness capital). Unfortunately, distrust capital is transferred and multiplied very easily and fast. These two spirals of trust and distrust (mistrust) are very difficult to break. Merton (1996, p. 318–336) formulated the so-called "Matthew Principle" based on the Gospel according to St. Matthew, saying that the rich will be richer and the poor will be poorer.

Every organization consists of nine essential elements:

1. Adoption
2. Understanding
3. Teaching
4. Solving
5. Thinking
6. Communication
7. Values
8. Behavior
9. Knowledge (Grudzewski and Hejduk 1997, p. 3–4).

But, at the same time, every organization is founded on **trust** (Figure 1.6). These nine elements enhanced with trust determine the organizational development level. They constitute the network of mutual feedback. In order to denominate abilities and functionality of this network, we need to analyze the following four cycles:

1. **Awareness cycle**—combining knowledge with adoption and understanding
2. **Adaptation cycle**—determined by teaching and solving
3. **Innovation cycle**—defined by thinking and communication
4. **Realization cycle**—conducted by values and behaviors (Grudzewski and Hejduk 1997, p. 4–8).

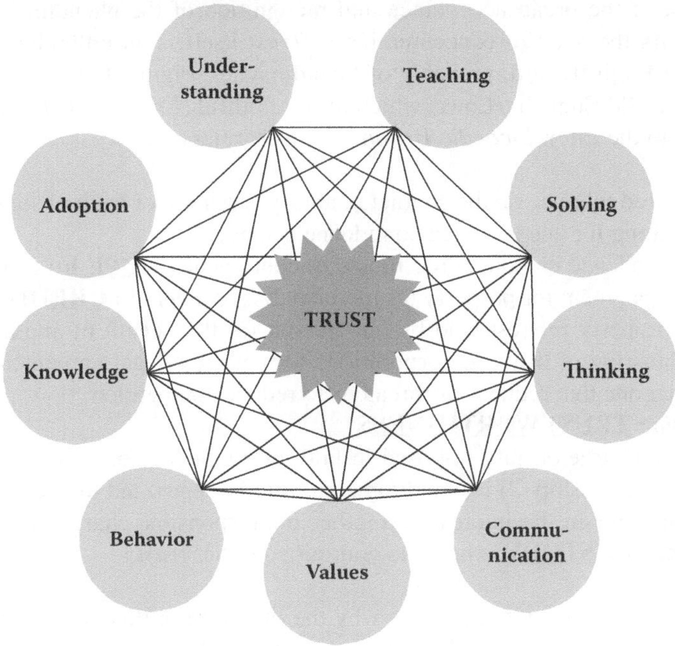

FIGURE 1.6 Network and trust.

Trust predisposes to the role of the strategic supply in organization:

1. **Value test**—trust helps organizations adapt to dynamic changes in turbulent environments
2. **Rarity test**—high levels of trust are nontangible assets in only some organizations
3. **Ownership test**—interpersonal trust is seen as specific "employee's corporation" because it can be observed in a certain employee
4. **Imitation test**—trust is highly resistant to imitation or automatic copying (it is very difficult to imitate)
5. **Resistance test**—when creating trust with time, there is a tendency to trust growth
6. **Substitution test**—trust cannot be replaced with other utilitarian value because it is the driving force for new culture norms
7. **Competitiveness test**—trust can form the basis for the highly competitive action strategy
8. **Formalization test**—trust cannot be created through administrative regulations and codified organization rules
9. **Organization test**—trust practically contains all aspects of enterprises' functioning.

In light of the broad acceptance and recognition of the phenomenon of trust, there appears the need to conceptualize it. *Trust* itself is an umbrella term. Trust is a Middle English word, probably of Scandinavian origin, akin to the Old Norse *traust* and to Old English *trEowe,* which meant faithful (*Encyclopaedia Britannica*). According to the *Encyclopaedia Britannica,* trust refers to:

1 a : assured reliance on the character, ability, strength, or truth of someone or something **b :** one in which confidence is placed
2 a : dependence on something future or contingent **: HOPE b :** reliance on future payment for property (as merchandise) delivered **: CREDIT**
3 a : a property interest held by one person for the benefit of another **b :** a combination of firms or corporations formed by a legal agreement; *especially* **:** one that reduces or threatens to reduce competition
4 *archaic* **: TRUSTWORTHINESS**
5 a (1) **:** a charge or duty imposed in faith or confidence or as a condition of some relationship (2) **:** something committed or entrusted to one to be used or cared for in the interest of another **b :** responsible charge or office **c : CARE, CUSTODY** <the child committed to her *trust*>

Trust is a very old notion; that is why the literature in this field is full of many authors studying its nature and essence. As a result, there are many definitions of *trust* (Table 1.1).

The definitions presented in Table 1.1 are a result of the interests from various disciplines, such as psychology, management, marketing, organizational behavior, public relations, and information systems. The most common characteristics used in trust definitions are **benevolence, integrity, competence,** and **predictability.** *Benevolence* means care and motivation to act for somebody else's goodness, and it is something totally opposite to opportunistic behavior (Holmes 1991). *Integrity* characterizes concluding an agreement with good will, telling truth, and keeping promises (Bromiley and Cummings 1995). *Competence* defines skills or capabilities necessary to do what we need to do (Barber 1983). *Predictability* addresses the actions of partner (wanted as well as unwanted) that are consistent enough to predict future situations (Gabarro 1978). By analyzing trust definitions available in the trust literature, we can examine several very important issues associated with trust. Schumacher (2003) clarified the following key characteristics of trust, namely:

- Trust exists in an environment of *uncertainty and risk.* We can never be entirely sure what the final result will be. We should be prepared for contingencies depending on other companies, institutions, and persons. Therefore, we can easily conclude that if partners have done a deal in the conditions of total certainty, trust would not be necessary and the concept of modeling trust would be trivial.
- Trust suggests that a partner is *exposed to risk.* The size of potential loss caused by untrustworthy behavior is usually significantly bigger than profits expected from honest activity. However, at the same time, bigger risk

TABLE 1.1
Review of Trust Definitions According to Various Disciplines

		Psychology and Sociology
Rotter	1967 and 1980	A generalized expectancy held by an individual that the word, promise, or oral or written statement of another individual or group can be relied upon.
Griffin	1967	The reliance upon the behavior of a person in order to achieve a desired but uncertain objective in a risky situation.
Larzelere and Huston	1980	Trust exists to the extent that a person believes another person (or persons) to be benevolent and honest.
Cook and Wall	1980	The extent to which one is willing to ascribe good intentions to and have confidence in the words and actions of other people.
Pruitt and Carnevale	1993	The expectation that the other party will cooperate in the future.
Sztompka	2007	"Bet" (belief and, based on it, activity) that future activities of other people or functioning of the devices or institutions will be advantageous to us.
		Management
Zand	1972	Actions that (a) increase one's vulnerability, (b) to another whose behavior is not under one's control, (c) in a situation in which the penalty (disutility) one suffers if the other abuses that vulnerability is greater than the benefit (utility) one gains if the other does not abuse that vulnerability.
Gulbertand and McDonough	1986	Trust pertains to whether one individual is able to value what another is up to and demonstrate respect for him or her, particularly when the individual's need and those of the person taking the action momentarily compete.
Gambetta	1988	When we say we trust someone or that someone is trustworthy, we implicitly mean that the probability that he will perform an action that is beneficial, or at least not detrimental, to us is high enough for us to consider engaging in some form of cooperation with him.
Mishra and Morrisey	1990	One party's willingness to be vulnerable to another party based on the belief that the latter party is (a) competent, (b) open, (c) concerned, and (d) reliable.
Carnevale and Weschler	1992	The expectation of ethical, fair, and nonthreatening behavior and concerns for the rights of others.
Sabel	1993	The mutual confidence that no party to an exchange will exploit the other's vulnerability.
Mayer, Davis, and Schoorman	1995	The willingness of a party to be vulnerable to the actions of another party based on the expectation that the other party will perform a particular action important to the trustor, irrespective of the ability to monitor or control that other party.
McAllister	1995	The extent to which a person is confident in, and willing to act on the basis of, the words, actions, and decisions of another.

(continued)

TABLE 1.1 (CONTINUED)
Review of Trust Definitions According to Various Disciplines

		Management (continued)
Fukuyama	1995	The expectation that arises within a community of regular, honest, and cooperative behavior, based on commonly shared norms on the part of other members of that community.
Rousseau, Sitkin, Burt, and Camerer	1998	Trust is a psychological state comprising the intention to accept vulnerability based on positive expectations of the intentions or behaviors of another.
Doney, Cannon, and Mullen	1998	A willingness to rely on another party and to take action in circumstances where such action makes one vulnerable to the other party.
Bhattacharya, Devinney, and Pillutla	1998	Trust is an expectancy of positive (or nonnegative) outcomes that one can receive based on the expected action of another party in an interaction characterized by uncertainty.
Shockley-Zalabak, Ellis, Cesaria	2003	An organization's willingness, based on its culture and communication behaviors in relationships and transactions, to be appropriately vulnerable based on the belief that another individual, group, or organization is competent, open and honest, concerned, reliable, and identified with common goals, norms, and values.
		Marketing
Anderson and Narus	1990	A firm's belief that another company will perform actions that will result in positive outcomes for the firm as well as not take unexpected actions that result in negative outcomes.
Moorman, Deshpande, and Zaltman	1993	A willingness to rely on an exchange partner in whom one has confidence. It can be recognized as (1) belief, sentiment, or expectation or (2) behavioral intention that reflects relying on an exchange partner as well as containing sensitivity and uncertainty about the trustor.
Morgan and Hunt	1994	When one party has confidence in an exchange partner's reliability and integrity.
Delgado-Ballester, Munuera-Aleman, and Yague-Guillen	2003	Feeling of security held by a consumer in his or her interaction with the brand, based on perceptions that the brand is reliable and responsible for the interests and welfare of the consumer.
		Organizational Behavior
Zucker	1986	A set of expectations shared by all those involved in an exchange.
Blomqvist and Stahle	2000	An actor's expectation of the other party's competence, good will, and behavior.
Zaheer, McEvily, and Perrone	1998	The expectation that an actor (1) can be relied on to fulfill obligations, (2) will behave in a predictable manner, and (3) will act and negotiate fairly when the possibility for opportunism is present.
Sitkin and Roth	1993	A belief in a person's competence to perform a specific task under specific circumstances.

(continued)

TABLE 1.1 (CONTINUED)
Review of Trust Definitions According to Various Disciplines

		Public Relations
Broom, Casey, and Ritchey	1997	Predictability that one party will behave in a specific and desired way.
Heath, Seshandri, and Lee	1998	Perceived competence, objectivity, fairness, consistency, good will, confidence, and the prediction that others will act supportively rather than exploitatively.
Ledingham and Bruning	1998	A feeling that those in the relationship can rely on each other.
Grunig and Hon	1999	One party's level of confidence in and willingness to open oneself to the other party.
		Information Systems
Keen	1999	Trust in business relation. The definition contains risk and concentrates on relations connected with computers, telecommunication, creation of trust relationships (safety, honesty, consumers' rights, contracts, privacy, reputation, trademark, mutual interest).
Ratnasingam and Pavlou	2002	Subjective probability used by members of an organization to selectively assess whether a certain deal was done.

allows one to expect bigger profits. Therefore, some level of risk is always expected and simultaneously trust is wanted by the partners.

- Trust describes some level of *predictability*. It reflects the expected behaviors of people and hopes that a certain person will accomplish his task well enough that other following partners will consider involving him in any form of collaboration in the future. Predictable partners are an important condition to building trusting relationships and continued cooperation in the futuro.
- Trust exists in the conditions of *interdependence and mutuality*. The condition for collaboration and cooperation is the conviction that partners cannot accomplish their goals as a result of dependences among them. Nowadays, acting alone without any partners is impossible and collaborators are indispensable to each other. This situation tightens the relationships and enhances the need for trust.
- Trust, by its very nature, invokes the notion of *positive and good character*. Talking about trust, we think about the probability of gaining positive, targeted results. Trust is always combined with reliability, honesty, benevolence, openness, and care.

Trust is a complex phenomenon that influences many parties and their activities. The ability to trust is crucial but, according to Gambetta (1988, p. 221), it may be equally essential to be trusted by others. Generally speaking, trust is a controlled or directed relation between two units; one unit is called *trustor A*, and the second one is called *trustee B*. The trustor, as a matter of fact, needs to be a "thinking entity," whereas the trustee can be represented, for example, by a person, a physical object, or an abstract notion, such as *software*. The source of trust can be:

1. Familiarity resulting from certain repeated interactions between the partners or parties
2. Calculatedness constituting a subjective assessment of predominance of benefits over potential costs
3. Values believed by B norms help to increase positive belief/conviction about his honesty and good will.

Trust appears in the **interdependence** conditions, between trustor and trustee, bearing the hallmarks of **risk.** If both parties did a deal under the conditions of certainty, the conception of trust would be trivial. From this point of view, risk is not considered in terms of probability of positive or negative deviations from expected result, as it is defined in finances. Risk is seen as a subjective probability of negative consequences of actions taken in the conditions of uncertainty (Sztompka 2005, p. 326). Risk rises as the number of prospective partners grows because these partners are more and more varied, distant in space, less visible, and less controllable for us. At the same time, we cannot avoid this kind of risk in interactions with other people. Partners need to constantly start this game, take bets, and foresee different actions of other people.

Furthermore, a *trusting relationship* refers to an agreed scope of situations, embracing a defined purpose of activity and/or defined area of activity. Mutual trust can be found when two individuals trust the same object, which results in the fact that they both need to be the "thinking entity." Trust influences the attitudes and actions of the trustor and, at the same time, affects the trustee and other elements from their environments (for example, developing *reciprocal trust*).

Furthermore, a distinction between a context-independent trust (discerned as a reliability trust) and a context-dependent trust (discerned as a decision trust), is reconnoitered. Gambetta (1990) suggests the following definitions:

Definition 1.1 Reliability Trust Trust is the subjective probability by which an individual expects that another individual performs a given action on which its welfare depends.

In Definition 1.1, trust is primarily understood as the trustor's estimation of the trustee's reliability and includes the concept of dependence on the trustee. This reliability is considered in the sense of probability. This subjective probability reflects a subjective conviction that a certain occurrence will occur and frequently it results from epistemological interdeterminism. It can be noted that this definition does not take the context into account. However, trust can be more complex and made up of many different parts than Gambetta's (1990) definition indicates. Falcone and Castelfranchi (2001) assert that having high (reliability) trust in a person in general is not sufficient enough to decide to enter into a relation of interplay. For instance, it is possible that the value of the damage *per se* (in case of failure) is too high to choose a given decision branch, and this independently is spawned by either the probability of the failure (even if it is very low) or from the possible payoff (even if it is very high). In other words, that danger might seem to the agent as an intolerable risk. It is illustrated quite well in an example from Jøsang, Keser, and Dimitrakos (2005). We consider an example of a person who distrusts an old rope for climbing from the third floor of a house during a fire exercise. Imagine now that the same person is trapped in a real fire in the same house and that the only escape is to climb from the third floor window with the same old rope. In a real fire, most people would trust the rope (although the reliability trust in the rope is the same in both presented situations), but the decision about trust changes along with the values of the utility function associated with the possible courses of action. This example describes the concept of trust seen within certain, met situations or circumstances (decision trust [see below]).

The definition of **context ø dependent** (also known as *decision trust* or *situation-dependent trust*) is as follows (Jøsang, Keser, and Dimitrakos 2005):

Definition 1.2 Context ø dependent trust Trust is the extent to which a given party is willing to depend on something or somebody in a given situation with a feeling of relative security, even though negative consequences are possible.

In other words, A trusts B in certain context ø. Definition 1.2, apart from the notion of dependence and reliability touched on in definition 1.1, also aggregates such elements as usefulness (possible results), environmental factors (contracts, safety mechanisms, and so on), and the attitude toward risk (acceptance of risk, aversion to risk). In most cases, trust is selective; it refers only to chosen objects or people (e.g., I trust my wife and children, but I don't trust my neighbors; I trust my employer, but I don't trust my workmate; I trust banks, but I don't trust broker offices). Therefore, we can introduce **trust balance** (Sztompka 2005, p. 312)—that is, the range of objects and people who entity A trusts and to what extent entity A trusts them (level of trust, scope of trust, and so forth).

Trust Principles

In literature from this subject, there can be traced trust principles. Below are presented the most common and often quoted trust principles.

Principle 1.1: Trust transitivity

In the case of trust, one's own experience of the relying party in relations with a trustee is more important than any positive or negative reputation. Nevertheless, as a result of personal inexperience, trust based on secondhand information plays a significant role and usually trust originates from this kind of information, not from past experience. Now we can outline a very important badge of trust, its **transitivity.** The idea behind trust transitivity is that when, for instance A trusts B and B trusts C, then A may derive a measure of trust in C from B's referral, as it is drawn below in Figure 1.7. Here, Anne trusts Bill, and Bill trusts Cindy. In this situation, there appear some possibilities for people from outside, third parties, to certify trust of certain entities in certain contexts.

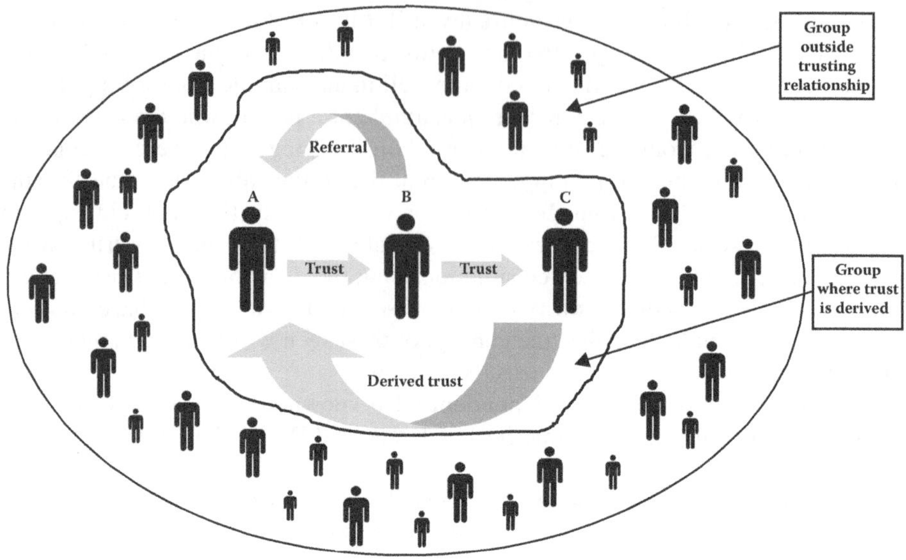

FIGURE 1.7 Trust transitivity principle according to Jøsang, Keser, and Dimitrakos (2005). *Source:* Adapted from Jøsang, A., Keser C., and Dimitrakos T. 2005. Can we manage trust? In *Trust Management: Third International Conference, iTrust 2005, Paris, France, May 23–26, 2005, Proceedings (Lecture Notes in Computer Science)*, P. Herrmann, V. Issarny, and S. Shiu (eds.), 93–107. Berlin/Heidelberg: Springer.

Principle 1.2: Trust is a function of risk perception

Trust is a belief or an expectation that we can place confidence in somebody and this person will not take advantage of our frailty (Geyskens, Steenkamp, Scheer, and Kumar 1996). In this way, trust reflects a subjective conviction in conditions of uncertainty that the other party will act in a certain way and not in any other way as well as that we have only a weak knowledge of the observator about *status quo*.

Principle 1.3: Trust is determined by time

Trust is seen as a dynamic phenomenon (Heimer 2001), changing as a consequence of collaboration processes *inter alia* as a result of accumulated experiences. However, the way experiences are being collected is changing in the process of time because of the variety of initial actions performed by parties as a result of simultaneous learning and forgetting. The mathematical representation of the process of experience accumulating is given by the formula below (Luna-Reyes, Cresswell, and Richardson 2004):

$$\frac{d}{d_t}(A's\ Knowledge\ about\ B) = (Learning - Forgetting) \tag{1.3.1}$$

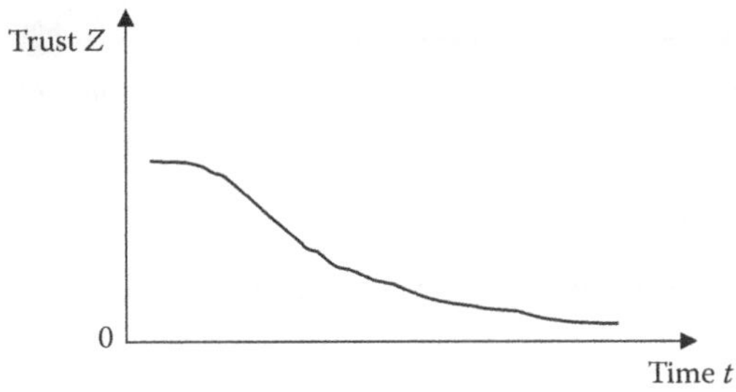

FIGURE 1.8 Dynamics (decay) of trust in case of lack of renewing interactions.

Dynamics of trust changes has one crucial temporary characteristic. In the process of time, when there are no interactions renewing knowledge of A about B due to a general tendency to forget, trust decreases with time function (Figure 1.8). At the very beginning, in the first period of relatively short time, trust does not change; then it declines dramatically in time as a result of the forgetting curve running to almost neutral attitude; and then it starts falling slightly. The forgetting curve implies that humans tend to halve their memory of newly learned things from interaction in a matter of time unless they review the learned issues.

Trust is often characterized as a trajectory of dependences. A trajectory of dependences is a course of behaviors in creating trust. This issue has not been well investigated so far. Its main characteristic is that small, accidental events in the early history of a system determine the final state even if all states are equally probable at the beginning (Sterman 2000). On the basis of theoretical discussions and empirical observations, trajectory of trust is recognized as nonlinear. Assuming that to be true, the extreme values of trust and mistrust are more probable (Rousseau, et al. 1998). In addition, various types of trust and different levels of its intensification are connected with subsequent phases in business relations (Zucker 1986). Seeing that, Lewicki and Bunker (1998) built a model for interpersonal trust based on three foundations: deterrence-based trust, knowledge-based trust, and identification-based trust. Trust develops with time from the lowest level based on a deterrence factor through knowledge-based trust to identification-based trust, which is characterized by the highest level of trust.

Principle 1.4: Trust can be quantified

There is no complete trust. To be more precise, degree of trust is a function of subjective certainty referring to distinct beliefs and expectations. If trust degree is a measure normalized in a range [0,1], then we can describe the degree of trust as:

$$Z(A,B,\phi) \rightarrow [0,1]$$

where value *0* means no trust, and *1* equals complete trust.

Defining "the strength of trust" is useful in making rational decisions by A referring to relying on B. On the basis of common acknowledgement of trust measurement, there are methods of estimating it. Generally speaking, measures of trust

FIGURE 1.9 Trust degree.

expressed numerically are based on numerical components (see "Methodology of Trust Measurement").

Principle 1.5: Culture influences trust

The fundamental basis of trust differs depending on organizational culture and national culture. This relationship is the subject of subsequent chapters.

Principle 1.6: Formal and social tools for control are significant for trust development

Formal tools used for control contain systems of codified principles, targets, procedures, and regulations defining expected behavioral patterns (Das and Teng 1998). Social control uses cultural and organizational values and norms in order to encourage certain behaviors. For alliances, we can adapt a form of a get-together, mutual interactions as well as training (Das and Teng 1998).

Types of Trust in Business Relationships

For better understanding of the trust essence in business relationships, we can define possible forms of trust using the basis for trust and explaining how it changes during the collaboration. Below are outlined trust typologies in business relationships according to certain authors (Ratnasingam 2003).

Dwyer, Schurr, and Oh (1987) propose a five-phase development of relations between business partners:

- **Phase 1. Consciousness** appears when business partner A initially learns about business partner B using a study of his feasibility. In e-commerce, the partners' selection is usually conducted in order to find the best partner.
- **Phase 2. Recognition** is done when a business partner seeks and tries to make a deal. Benefits and risks of a business relationship are investigated. This phase consists of five stages: attraction, communication, negotiation, development and use of power, evolution of norms, and growth of expectations.
- **Phase 3. Expansion** is when business partners permanently aim to increase gained benefits. This allows each partner to get satisfaction from the received results. For instance, if one business partner accomplishes the expected deal or transaction and the obligation has been fulfilled perfectly, attractiveness of this business partner increases.
- **Phase 4. Commitment** can be in the form of a visible or hidden obligation that secures the continuity of the business relationship. In bilateral relationships, commitment is based on structures with control or common values. Commitment is combined with solidarity (conviction that the future

collaboration will be effective) and cohesion as well as durability and consequence of actions.
- **Phase 5. Durability** of business relation in time refers to the fact that business partners are aware of the benefits resulting from the continuation of the business relationship.

Gabarro (1978) alleges that social and *stricte* business relationships evolve in time and can differ in terms of their durability, mutuality, and effectiveness. He performed research on newly appointed chairmen, observing how they developed their relationships with their key subordinates. The results of his research suggest that trust is present in the process of open and coherent communication. A process recognized as "interpersonal contract" is determined by the set of mutual expectations, connected with them results, roles, trust, and influences. Gabarro (1978) discerned four stages in development of newly engaged working relationships:

- **Stage 1, Orientation process**—when employees formulate their first opinions about other employees
- **Stage 2, Recognition process**—when the initial expectations become clearer; lasts several months
- **Stage 3, Trust testing in relationships**—when an interpersonal contract is created that defines the limits of each employee's influence; lasts from 6 months to 1 year
- **Stage 4, Grounding the interpersonal contract's durability**—durability may also be modified a little bit.

Gabarro (1978) characterizes trust via various elements, such as predictability, communication openness, and creditability (the degree to which a certain person is convinced that another person will not start any unwanted and arbitrary actions).

Donney and Cannon (1997) insightfully studied relationships between buyers and suppliers. On this basis, they identified the following five processes in trust development:

1. **Calculation process,** when the parties are calculating costs and benefits and costs of interactions with the business partner.
2. **Process of skills' assessment,** containing the second party's determination to meet its obligations via focusing mainly on one aspect of trust—creditability.
3. **Process for predicting,** during which each party uses its skills, built upon previous interactions, to predict the business partner's behaviors. Lewicki and Bunker (1996) describe predictability as an element of trust that needs not only to be repeated but requires consistent interactions.
4. **Process based on target,** which appears in situations when a trustor interprets words and behaviors and assesses the intentions. Trust emerges from interpretation and assessment of business partners' motives.
5. **Transfer process,** including the pattern used to gain trust using reputation and recommendations. In this process, we use the term "third party" of

hitherto prevailing business partner as a warrant to believe that the third party is worthy of trust.

Ring and Van de Ven (1994) prepared a framework describing how intra-organizational relations are developed and finished. They believed that intra-organizational relations characterized as cooperation consist of repeated sequences of negotiations, commitment, and fulfillment stages. During the phase of negotiations and drawing up an agreement, the business partners define their mutual (not individual) expectations concerning their motivation, prospective investments, and noticed uncertainties. The stage of commitment aims to fulfill the expectations of the parties. They reach an agreement on obligations and principles needed for future actions. Structure of control is defined in this stage, leading to formally agreed relationships or informal agreements in the form of a psychological contract between the parties. In the stage of fulfillment, the parties fulfill their obligations according to the agreed rules. Ring and Van de Ven highlighted in their opinions the development of intra-organizational relations in the context of trust.

Mayer, Davis, and Schoorman (1995) recognized three elements defining trust. They are as follows:

1. Skills, competences, and characteristic traits of business partners, which allow one party to have influence on the other party.
2. Benevolence or willingness to do good to trustor despite the possibility of egocentric motivation to act for profit. Benevolence means that the trustee feels a special attachment to the trustor.
3. The trustor's perception of the trustee's honesty appears when the trustee follows the set of principles determining his reliability and predictability (if his actions are consequent).

Mayer, Davis, and Schoorman (1995) added to the trust definition various behaviors observed in different phases of a business relation.

According to Mishra (1996), trust is the willingness of one party to be exposed to unwanted actions from the other party based on the assumption that the other party is:

1. competent
2. reliable
3. open and caring.

These three dimensions strengthen trust and stimulate interest in the remaining business relationship with the partner. These dimensions are described as follows:

- **Competences**—identifying the skills of the business partner to proper data interpretation and consequently the ability to make the right decisions. Consistent, competent behavior builds the credibility of the business partner. The higher credibility, the higher confidence and willingness to trust the party and the higher value of the business relationship.

- **Openness**—the skill to act on the basis of integrity. Openness implies that the partner is willing to accept new ideas and share information rather than control the situation and hide information. This demonstration reduces the probability of opportunistic behaviors from the partner. Openness to changes improves the effectiveness that promotes mutual interest to continue the collaboration. Continuity increases trust that the partner will not behave in an opportunistic way (e.g., by using information inappropriately or lying).
- **Care**—implies that the partner will act in a way that profits the other party. Care leads to the belief that the partner will not make a use from the trustor.
- **Reliability**— correspondence between declarations and actions. Reliability determines the level up to which a trustor might rely on a trustee and make sure the partner can keep promises. Reliability assists collaboration, which then strengthens trust. Cooperation is the set of supplementary coordinated organizational actions and cooperating relations aimed to gain mutual profits rooted in certain mutuality in the future.

Lewicki and Bunker (1996) developed Shapiro's model (1992), adding **psychological perspective.** They claimed that trust development goes through trust phases based on a deterrence factor, knowledge, and identification. They suggested that each type of relationship (romantic, manager–employee, between business partners) is connected with a course of mentioned trust phases. Their model contains the following types of trust:

- **Deterrence-based trust,** or **calculus-based trust,** emphasizes the logic of business partners' behaviors. The fear of punishment is seen as a negative factor; at the same time, economic benefits and awards are recognized as something positive. Lewicki and Bunker (1996) suggest that trust based on deterrence factor is not premised on fear of punishment for trust peculation but on award for fulfilling one's obligations. Dilemma of course of actions selection in relationship based on the balance of benefits is illustrated in Figure 1.10.
- **Knowledge-based trust** originates from predictability of behaviors coming from previous experiences in cooperation with the partner (trustee).
- **Identification-based trust** is built on empathy, care, and common values shared by partners. It focuses more on common tasks than on individual preferences.

The model of trust proposed by Lewicki and Bunker (1996) presupposes that deterrence-based trust depends on the belief that there is a reliable risk of punishment when the collaboration fails. So trust is here the result of an economical account in a certain situation. Formally, it can be presented as a function of B's attractiveness for A and the attitude of A toward the risk (Luna-Reyes, Cresswell, and Richardson 2004), as shown here:

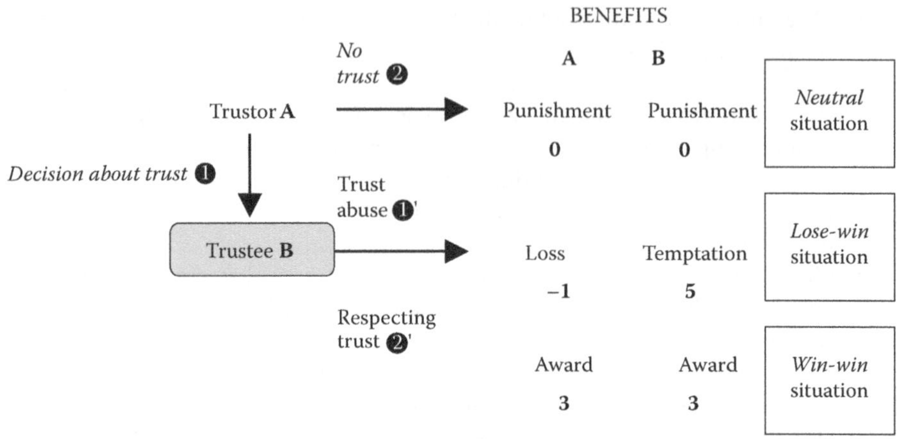

FIGURE 1.10 Trust dilemma.

$$Calculative\ Risk = f\left(\frac{\dfrac{Attractivenss\ of\ B\ to\ A}{Risk}}{Attitude\ \ of\ \ A\ to\ Risk}\right) \tag{1.3.3}$$

The value of index of B's attractiveness for A divided by risk in a certain context lower than 1 means that A accepts the situation when the risk is bigger than the attractiveness (a *risk hunter*). When this index is higher than 1, it means that A prefers a situation in which attractiveness is higher than risk (a *risk escaper*).

Knowledge-based trust assumes that human aptitudes are well known and can be predicted quite well. This model agrees with the opinion about the rational premises of human choices. Finally, identification-based trust appears when a business partner fulfills the expectations of the other partner in a way that allows them to reach common goals.

McAllister (1995) provided insight into two basic forms of interpersonal trust: **cognition-based trust** (also called **rational trust**) and **affect-based trust** (sometimes called **emotional trust**). Cognition-based trust originates from individual belief about reliability and confidence, whereas affect-based trust is also built on mutual interpersonal care and feeling of attachment and good will. Elements building cognition-based trust are reliability, competences, honesty, and consequence of business partners. Affect-based trust includes such building blocks as faith, care, openness, support, and information sharing among parties. Cognition- and affect-based trust suggest the selection who we can trust and under what conditions. The selection is based on certain premises that are the trust representatives (Lewis and Weigert 1985). To recapitulate, research done by McAllister brought into the discussion of various types of trust two new types of behavior—cognitive-rational *versus* affect-emotional. Moreover, he discovered that development of cognition-based trust is positively correlated with forming affect-based trust and it is before its development.

According to Barney and Hansen (1994), trust is classified into three groups, namely:

1. **Weak trust**—characterizing partners who believe that the other party will not make use of their vulnerability. Partners are not obligated by any contractual or other form of mechanisms of control. Weak trust appears when there are limited possibilities for opportunistic behaviors. (Das and Teng 1996, Parkhe 1998, Sitkin and Pablo 1992).
2. **Semi-strong trust**—hanged on control mechanisms like reputation, protective contracts just in case there will be any opportunistic behaviors. Semi-strong trust appears when there exist significant areas of the partners' sensibility to an attack. This type of trust comes from the protection included in the contracts. Semi-strong trust is especially beneficial when other methods of control and development of suitable skills are too costly to be started.
3. **Strong trust,** also called **principled trust**—comes as the answer for the set of acquired norms and principles directing the business partners' behaviors. Trust of this type depends on the existence of certain control mechanisms and the difficulty in imitating them.

Identifying theses types of trust, Barney and Hansen showed the relations between certain levels of trust and different control mechanisms.

Sako (1998) recognized three types of trust:

1. **Contractual trust** depends on skills to obey the agreement points by business partner. It covers mutual norms, honesty, keeping promises, mutual understanding of principles for professional behavior, and technical and managerial standards.
2. **Competence trust** is connected with the tendency toward keeping promises. These are abilities of a business partner to take active part in business operations.
3. **Good will trust** emanates from the commitment of business partners and mutual benefits resulting from this commitment. Good will trust can appear only when there is consensus about principles of integrity. This type of trust is believed to be the most cooperative (helping collaboration via trials aiming to satisfy the needs of partner) and highly assertive (taking into account deriving one's own satisfaction without depletion of the partner's benefits).

Zucker (1986) postulated three sources of trust:

1. **Process-based trust** originates from certain social experiences or economic transactions. There is an assumption that the desire for future collaboration is strongly connected with constancy of personal relations. It coincides with the point of view of Granovetter (1985) and Fukuyama (1995), who suggest that there is a social ground for economic actions.
2. **Characteristic-based trust** refers to personality traits, such as age, sex, and membership in certain ethnic groups or social systems. Exchange is limited to these people who agree on a common accepted system, origins, and expectations.
3. **Institutional-based trust** goes beyond certain experiences in collaboration and does not depend on the exchange partner. Sources of this trust are traditions, certificates, licenses, and trademarks.

Dyer and Chu (2000) proposed a trust model composed of three items: reliability, honesty, and reputation. Trust is based on conviction, which develops in situations when trustworthy business partners during the relation are:

1. Dependable in their behavior and fulfill their obligations
2. Adapting (because market conditions are changing) and can be seen as honest by partners
3. Not using eventual opportunities against the business partner.

Bratnicki and Strużyna (2001) tried to synthetically unite the powers creating trust. As a result, they presented five dimensions of trust:

1. **Personal trust**—defined by structure of personality. It concentrates on inborn or gained personality traits of an organization's people (e.g., emphatic ability to understand other peoples' situations, sensitivity, responsibility, emotional restraint).
2. **Calculus trust**—interpreted as trust based on rational assessment and the balance between possibilities and risks. It is created on the basis of prospected incomes and cost of continuing certain relations.
3. **Institutional trust**—supported by formal regulations. It emerges from the feeling of safety coming from using structural protection like guarantees, regulations, contracts, promises, and other formal forms. All above-mentioned elements influence the clearness and stability of the situation.
4. **Perceptive trust**—founded on perception and its essence are processes of mutual perception. It refers to cognitive traits of each individual and describes trust in the categories of observations of mutual behaviors, reactions, and conduct. Perception shapes trust as a derivative of self-esteem, justice in organization, or opinion about managers.
5. **Cumulative trust,** or **knowledge trust**—the most durable category of trust but it takes a lot of time to reach it. It is a result of gained and accumulated experience in mutual interactions.

TABLE 1.2
Summary of Trust in Business Relationships According to Different Authors

Types of Trust		Phase 1: Competence-Based Trust	Phase 2: Predictability-Based Trust	Phase 3: Trust Based on Good will
Source:		**Economic Calculus**	**Familiarity**	**Values**
Author(s)	**Year**			
Gabarro	1978	Character Role competences	Judgment	Motives/Intentions
Mayer, Davis, and Schoorman	1995	Ability	Integrity	Benevolence
Mishra	1996	Competence	Reliability	Openness Care Concern
Lewicki and Bunker	1996	Deterrence/Calculus	Knowledge	Identification
McAllister	1995	Cognitive	Cognitive → Affective	Affective
Barney and Hansen	1994	Weak form of trust	Semistrong form of trust	Strong form of trust
Sako	1992	Contractual	Competence	Good will
Zucker	1986	Process-based trust	Characteristics-based trust	Institutional-based trust
Dyer and Chu	2000	Reliability	Fairness	Good will
Bratnicki and Strużyna	2001	Institutional trust	Cumulative trust	Perspective trust, personal trust

The presented analysis of trust and accompanying business behaviors clearly show the tendency of trust to develop in time and to be characterized by certain phases. Table 1.2 summarizes the mentioned types of trust with the all their phases.

After such a wide look at trust, **trust management** can be conceptualized in two ways:

- Firstly, as a process by which individual A becomes trustworthy to other individuals. This trust is a significant criterion of success and survival because it makes individuals to collaborate with individual A. It is not rare for people to try to build fake images of their own reliability for personal benefits, for example.
- Secondly, as a process of assessing the reliability of other individuals, which is as important as the process of building one's own reliability.

Taking together these two aspects we can formulate the following definition of trust management:

Definition 1.3 Trust management The activities of creating systems and methods that allow relying parties to:

1. Make assessments and decisions regarding the dependability of potential transactions involving risk
2. Allow players and system owners to increase and correctly represent the reliability of themselves and their systems.

We contend that trust management is one of the biggest novelties in contemporary organizations (Figure 1.11).

For trust management defined above, there are some **crucial assumptions** as follows:

1. Trust and suspicion are often well founded because, in competing organizations, which often stay in the scope of unforeseeable political influence, the cost of trusting inappropriate object can have dramatic effects.
2. In potentially uncertain, dangerous, and risky environments, we need to know very well who we can and who we cannot trust and, then, in which circumstances we can do what. Moreover, it is essential to see the necessity of being aware of potential dangers.

The essence of trust management is not to trust but to decide to what extent we can trust and how to develop and create trust relationships. Adequate for this situation is a belief of the philosopher Thoreau that "we can trust without a risk more often than we think it is possible" (Kożusznik 2005, p. 142), which stands as the motto for this book.

When in a situation of trust deficit, which causes a climate of resignation and consternation and makes it hard to concentrate on agreed targets, a man naturally tends to limit mentioned unfavorable climate. This tendency is conducted in the process of **uncertainty reduction** and creating substitutes of trust (Sztompka 1998, 2007). These equivalents of trust are as follows:

1. **Providentialism**—meaning supernatural forces, fortune, God, or metaphysical forces. We believe that these are sources of certainty in the world and they take over control when a situation is getting worse. However, it leads to disastrous results, passivity and stagnation.
2. **Corruption**—predicting and having control over other people's actions using bribes, mutual favors, favoritism, and so on. It gives the illusion of predictability and order. In fact, corruption is creating a cynical world of mutual manipulations and exploitations.
3. **Pressure**—allows direct and unconditional control over other people's actions. It leads to excessive vigilance, taking over personal control. It results also in, for example, numerous private securities, whole communities in a certain district living around one or more streets protected

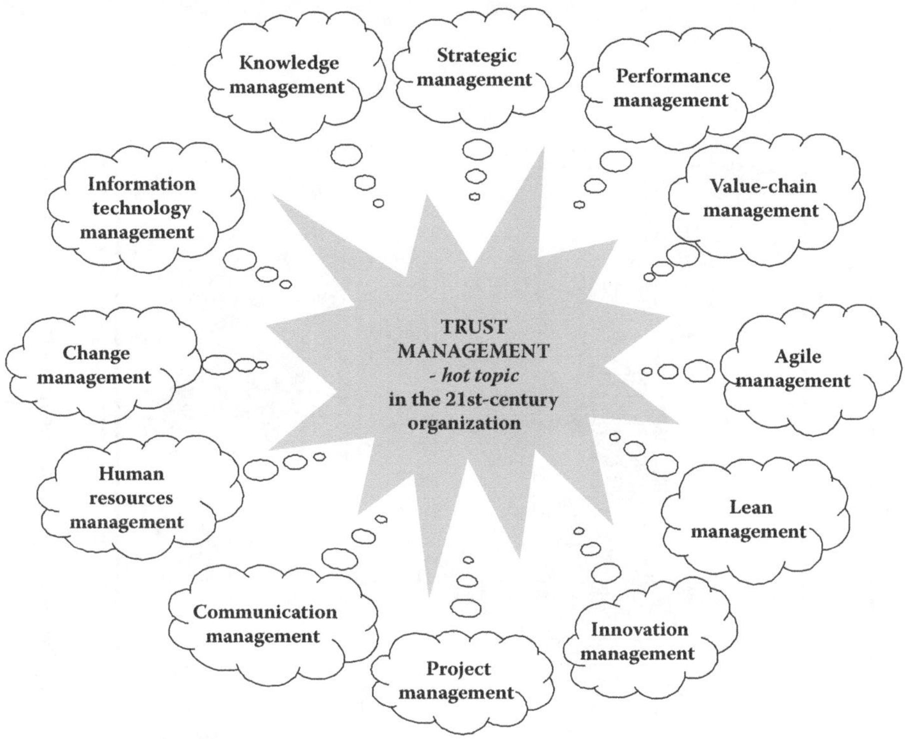

FIGURE 1.11 Trust management as a novelty.

by special guards, and an increasing number of guns owned by private people.

4. **Isolation (ghettoization)**—separating certain ethnic, religious, or professional groups from unfavorable and threatening their environments. People detach themselves from the outside world, reducing to some extent its complexity and uncertainty. This behavior can be observed among immigrants who do not assimilate with the citizens in the country to which they came to live their new lives. They continue to cultivate their own traditions, religion, language, and habits and live in closed groups with others immigrants with the same origins.
5. **Paternalism**—a strong authority, leadership figure, or "quasi father" (like *Fürher* or *Il Duce*) who assures missing order, hierarchical harmony, and predictability of the group's life.
6. **Externalization**—transferring trust to other external groups or communities together with a tendency to idealize procedure of operations, organizational solutions, or total quality of an institution of targeted group. There is a tendency to idealize foreign trustees. It is caused by the geographic distance, selected available relations in media, and lack of directly observed evidences against foreigners. Therefore, we believe that such international organizations as the North Atlantic Treaty Organization and the European Union will help us with our various problems.

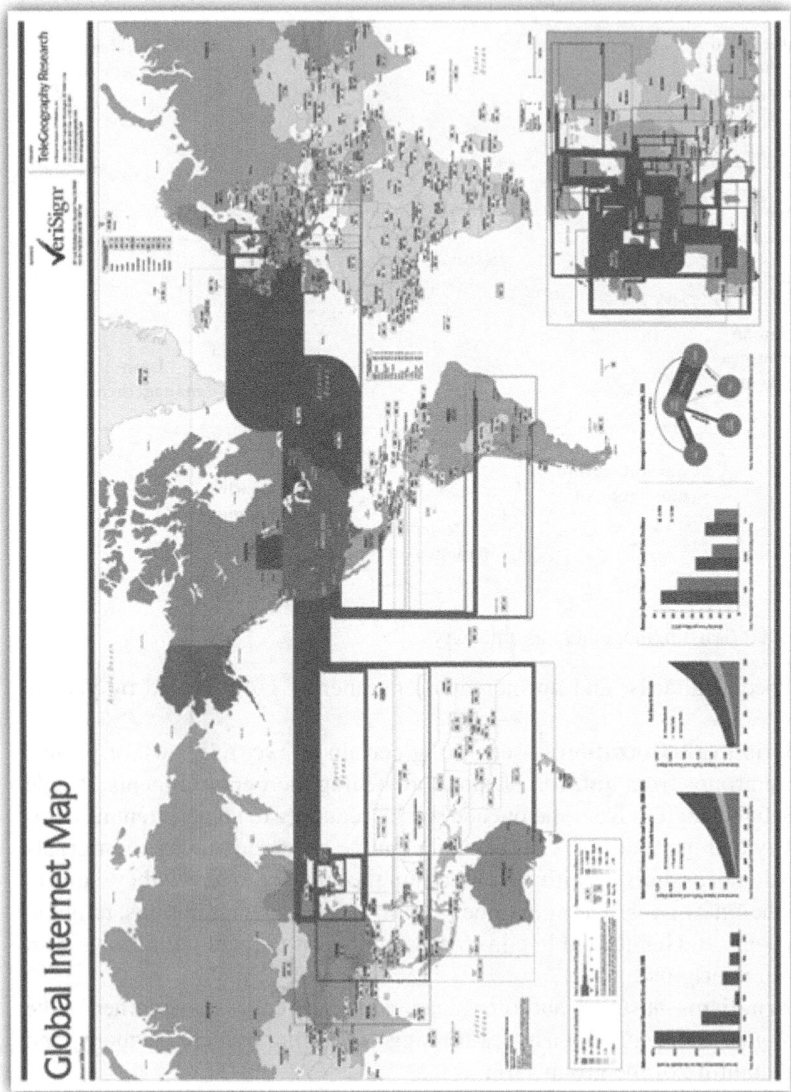

FIGURE 1.12 Map of global Internet, February 2006. *Source:* http://www.telegeography.com/products/map_internet/index.php.

7. **Excessive assistance of legal institutions**—when businessmen do not trust each other, they need formal means to secure the transaction (for instance, detailed contracts, bank guarantees, employing a certifying notary, and tendency to sue).

The role of trust in the contemporary world can be explained and revealed in analysis of the Internet teleinformation flow all over the world. Figure 1.12 shows a Global Internet Map with international teleinformation routes of at least 5 Gbps of aggregated capacity. It represents Internet bandwidth connected across international borders to each country. Domestic routes are omitted. One may observe that the map of Internet traffic looks more or less like the maps of maritime and land transport in the seventeenth century. Thus, it can be asserted that new technologies do not solve the problem, which does limit the number of new establishments of potential successful collaborations between parties relatively "strange" to each other. Most decisions about partners' selections are based on previous experience with already-known parties. As a result, decision-makers prefer to maintain the status quo and stay with already-known partners. For this reason, a number of potential successful collaborations that could bring new added value never occur. Practical implementation of trust management is expected to help overcome this situation of uncertainty in collaboration. Hence, trust management is a challenge for management science in the twenty-first century, particularly when nowadays virtualization has become a new paradigm.

REFERENCES

Akerlof, G. A. 1970. The market for "lemons": Quality uncertainty and the market mechanism. *Quarterly Journal of Economics*, 84:488–500.

Arrow, K. J. 1974. *The limits of organization*. New York: W. W. Norton and Company.

Barber, B. 1983. *The logic and limits of trust*. New Rutgers University Press, Brunswick, NJ.

Bratnicki, M. and Strużyna, J. (eds.). 2001. *Przedsiębiorczość i kapitał intelektualny*. Katowice: Wydawnictwo AE w Katowicach.

Bromiley, P., and Cummings, L. L. 1995. Transactions costs in organizations with trust. In *Research on negotiations in organizations*, R. Bies, B. Sheppard, and R. Lewicki (eds.), Vol. 5. Greenwich, CT: JAI, 219–247.

Ciancutti, A., and Steding, T. L. 2000. *Built on trust: Gaining competitive advantage in any organization*. Chicago: Contemporary Books.

Coleman, J. 1988. Social capital in the creation of human capital. *American Journal of Sociology*, 94:95–120.

Collis, D. J., and Montgomery, C. A. 1997. *Corporate strategy: Resources and scope of the firm*. London: McGraw-Hill/Irvin.

Das, T. K., and Teng, B.-S. 1998. Between trust and control: Developing confidence in partner cooperation in alliances. *Academy of Management Review*, 23(3), 491–512.

Dasgupta, P. 1988. Trust as a commodity. In *Trust: Making and Breaking Cooperative Relations*, D. Gambetta (ed.). New York: Basil Blackwell, Inc.

Earley, P. C. 1986. Trust, perceived importance of praise and criticism, and work performance: An examination of feedback in the United States and England. *Journal of Management*, 12(4), 457–473.

Encyclopaedia Britannica online. http://www.britannica.com/ (accessed November 13, 2006).

Falcone, R., and Castelfranchi, C. 2001. Social trust: A cognitive approach. In *Trust and deception in virtual societies,* C. Castelfranchi and Y. H. Tan, 55–99. Netherlands: Kluwer Academic Publishers.

Filek, J. 2003. O zaufaniu i pseudozaufaniu w przestrzeni gospodarczej. *Prakseologia,* 143, 59–72.

Friedman, D. D. 1986. *Price theory: An intermediate text.* Cincinnati: South-Western Publishing.

Fukuyama, F. 1995. *Trust: The social virtues and the creation of prosperity.* New York: The Free Press.

Gabarro, J. J. 1978. The development of trust, influence, and expectations. In *Interpersonal behavior: Communication and Understanding In Relationships,* A. G. Athos and J. J. Gabarro (eds.), 290–303, Englewood Cliffs, NJ: Prentice-Hall.

Gambetta, D. 1990. Can we trust trust? In *Trust: Making and breaking cooperative relations,* D. Gambetta (ed.), 213–238. New York: Basil Blackwell, Inc.

Gambetta, D. (ed.). 1988. *Trust: Making and breaking cooperative relations.* New York: Basil Blackwell, Inc.

Geyskens, I., Steenkamp, J.-B. E. M., Scheer, L.K., and Kumar, N. 1996. The effects of trust and interdependence on relationship commitment: A trans-Atlantic study. *International Journal of Research in Marketing,* 13(4), 303–317.

Gibb, J. R. 1978. *Trust: A new view of personal and organizational development.* Los Angeles: The Guild of Tutors Press.

Grudzewski, W. M., and Hejduk, I. 1997. Koncepcje kreowania organizacji wirtualnej w przedsiębiorstwach, *Organizacja i kierowanie,* 4(90), 3–22.

Handy, C. 1995. Trust and virtual organisation. *Harvard Business Review,* 73(3), 40–50.

Harisalo, R., and Miettinen, E. 1995. *Luottamuspääoma [Trust capital].* Tampere, Finland: Tampere University Press.

Harisalo, R., and Miettinen, E. 1997. *Trust capital: The third force of entrepreneurship.* Tampere University, Department of Administrative Science, Tampere: University of Tampere.

Heimer, C. A. 2001. Solving the problem of trust. In *Trust in society,* K. S. Cook (ed.), 40–88. New York: Russell Sage Foundation.

Holmes, J. G. 1991. Trust and the appraisal process in close relationships. In *Advances in personal relationships,* Vol. 2, W. H. Jones and D. Perlman (eds.), 57–104. London: Jessica Kingsley Publishers.

Jøsang, A., Keser, C., and Dimitrakos, T. 2005. Can we manage trust? In *Trust management: Third international conference, iTrust 2005, Paris, France, May 23–26, 2005, Proceedings (Lecture Notes in Computer Science),* P. Herrmann, V. Issarny, and S. Shiu (eds.), 93–107. Berlin/Heidelberg: Springer.

Keen, P. G. W. 1999. Transforming intellectual property into intellectual capital: Competing in the trust economy. In *Capital for our time: The economic, legal, and management challenges of intellectual capital,* N. Imparato (ed.), 3–35. Stanford, CA: Hoover Institution Press.

Knack, S., and Keefer, P. 1997. Does social capital have an economic payoff? A cross-country investigation. *Quarterly Journal of Economics,* 112(4), 1251–1288.

Kouzes, J. M., and Posner, B. Z. 1993. *The leadership challenge.* San Francisco, CA: Jossey-Bass Publishers.

Kouzes, J. M. and Posner, B. Z. 1995. *Credibility: How leaders gain and lose it, why people demand it.* San Francisco: Jossey-Bass Publishers.

Kożusznik, B. 2005. Wpływ a zaufanie. In *Wpływ społeczny w organizacji.* Warszawa: Polskie Wydawnictwo Ekonomiczne.

Lane, C. 1998. Introduction: Theories and issues in the study of trust. In *Trust within and between organizations: Conceptual issues and empirical applications*, C. Lane and R. Bachmann (eds.). New York: Oxford University Press.

Lasswell, H. D. 1958. *Politics: Who gets what, when, and how*. New York: Meridian Books.

Lewicki, R. J., and Bunker, B. B. 1998. Communication and trust in global virtual teams. *Journal of Computer–Mediated Communication*, 3, 4.

Lin, N., Cook, K. S., and Burt, R. S. (eds.). 2001. *Social capital: Theory and research*. New York: Aldine De Gruyter.

Lipset, S. M., and Schneider, W. 1983. *The confidence gap: Business, labor, and government in the public mind*. New York: The Free Press.

Luhmann, N. 1979. *Trust and power*. New York: Wiley and Sons.

Luna-Reyes, L. F., Cresswell, A. M., and Richardson, G. P. 2004. Knowledge and the development of interpersonal trust: A dynamic model. In *Proceedings of the 37th Annual Hawaii International Conference on System Sciences: 5–8 January 2004: Big Island, Hawaii (Hawaii International Conference on System Sciences//Proceedings)*.

McAllister, D. J. 1995. Affect– and cognition–based trust as foundation for interpersonal cooperation in organisations. *Academy of Management Journal*, 38(1), 24–59.

Merton, R. K. 1996. *On social structure and science*, P. Sztompka (ed.). University of Chicago Press.

Misztal, B. I. 1996. *Trust in modern societies*. Oxford: Polity Press.

Putnam, R. 1995. Tuning in, tuning out: The strange disappearance of social capital in America. *Political Science and Politics*, 28(4), 664–683.

Putnam, R. D., Leonardi, R., and Nanetti, R. Y. 1993. *Making democracy work: Civic traditions in modern Italy*. Princeton, NJ: Princeton University Press.

Ratnasingam, P. 2003. *Inter-organizational trust in business-to-business e-commerce*, London: IRM Press.

Ratnasingam, P., and Pavlou, P. 2002. *Technology trust: The next value creator in B2B electronic commerce*. Information Resources Management Association Conference, Seattle, Washington, May 19th–22nd, p. 889–894.

Rousseau, D. M., Sitkin, S. B., Burt, R. S., and Camerer, C. 1998. Not so different after all: A cross-discipline view of trust, *Academy of Management Review*, 23, 393–404.

Schumacher, C. R. 2003. The Influence of Trust on SME Co-operation Structuring and Performance. Discussion paper. Faculty of Business, Auckland University of Technology, New Zealand.

Shapiro, S. P. 1987. The social control of impersonal trust. *American Journal of Sociology*, 93, 623–658.

Sterman, J. D. 2000. *Business dynamics: Systems thinking and modeling for a complex world*. Boston: Irwin/McGraw-Hill.

Sydow, J. 1998. Understanding the constitution of interorganizational trust. In *Trust within and between organizations: Conceptual issues and empirical applications*, C. Lane and R. Bachmann (eds.). New York: Oxford University Press.

Sztaudynger, J. 2003. Zaufanie i kapitał społeczny a wzrost gospodarczy. Wyniki badań ekonometrycznych. *Prakseologia*, 143, 231–245.

Sztompka, P. 1998. Zaufanie: podstawa relacji społecznych. In *Krytyczna teoria organizacji*. v. 3, Warszawa: Wyższa Szkoła Przedsiębiorczości i Zarządzania.

Sztompka, P. 2005. Kultura zaufania. In *Socjologia, Analiza społeczeństwa*. Warszawa: Wydawnictwo Znak.

Sztompka, P. 2007. *Zaufanie. Fundament społeczeństwa*. Kraków: Znak.

Szulczewski, G. 2003. Zaufanie z perspektywy pragmatyczno-transcedentalnej koncepcji porozumienia. *Prakseologia*, 143, 141–151.

Watson, M. L. 2005, *Can there be just one trust? A cross-disciplinary identification of trust definitions and measurement.* The Institute for Public Relations, www.instituteforpr.com.

Whitney, J. O. 1996. *The economics of trust: Liberating profits and restoring corporate vitality.* New York: McGraw-Hill.

World Bank. 2000. http://www.worldbank.org/poverty/scapital/index.htm (accessed November 10, 2006).

Zucker, L. G. 1986. Production of trust: Institutional sources of economic structure. In *Research in organizational behavior,* Vol. 8, B. M. Staw and L. L. Cummings (eds.), 53–111. Greenwich, CT: JAI Press Inc.

1.4 FACTORS AFFECTING GENERALIZED TRUST

In the section "Theoretical Approaches toward Trust," the phenomenon of *trust* was defined in certain contexts and with reference to definite subjects and objects. Trust is bound up with definite conditions. In the conditions of uncertainty, we can make an assumption what the partner will do (Sztompka 2005, p. 326). Trust reflects knowledge about a specific partner resulting from previous experiences. Trust in certain situations is not only a result of the evaluation of a certain person and circumstances, but it is also determined by a general tendency (propensity) of a specific person to trust other people, the general "predisposition to trust" (Rotter 1967) named **generalized trust.** In this section, determinants of general trust are explored. This is crucial for better understanding of decisions made about trust in definite situations.

General trust is a belief that people generally are trustworthy. In other words, it is "trust in strangers," who may differ from us. This trust is measured using the answer to the following question: In general, how much can you trust other people and do you have to be careful in relationships with others? Sztompka (2005, p. 326) defines general trust as a willingness to act on the basis of the expectation *a priori* that most people and institutions will act in a way that is beneficial for them. Uslaner (2003, p. 3) considers general trust as a belief that most people belong to the same moral community. The general trust according to Uslaner (2003, p. 1) is highly appreciated in the knowledge society, and this phenomenon is believed to promote cooperation very efficiently. Nations with a high level of general trust (e.g., Scandinavian countries) have more open markets, higher economic growth, better governments, and less corruption and, at the same time, they are said to be leaders in globalization processes. We can speak here courageously about **the culture of trust** (Sztompka 2005, p. 326), which is a general trust cutting to the heart of the whole community and treated as an obligatory rule of conduct in the concrete community. **Detailed trust** is the conviction of an individual that other individuals connected within a certain network by social dependences or the social groups are reliable. Detailed trust refers to family, friends, collaborators, the religious community, associations, and clubs to which an individual belongs. **Particular trust** is a trust in people or groups that are similar to us. In the upshot, people *au fond* have tendencies to perceive members of their own group significantly more favorably than members "of strange groups" (Kożusznik 2005, p. 139). The superiority of general trust over particular trust is issued from the greater stability of this first in time (Uslaner 2002, p. 68). It is more difficult to modify through experiences of the individual, whereas particular trust might change in a short period, when local circumstances alter. Uslaner (2002, p. 43) affirms that particular trust constitutes the life of the nearest environment of the individual for better but does not establish a better life for the whole society; only general trust is able to do so. We can trust in our family members and friends and simultaneously acknowledge "the external world" as the source of evil and trickishness, which lead to alienation. It is socially dysfunctional. "Trust in strangers" gives the possibility of overstepping well-known social groups and encourages risk-taking connected with entering new socioeconomic interactions, generating new possibilities. The culture of trust leads to open, spontaneous, creative activities; raises the level of mobilization and activity of units; and intensifies interactions. As a result, general trust outruns the limits of face-to-face interaction, allowing people who do not know each other personally to join together (Stolle 2001, p. 205). This fact has an essential meaning in the Information Age and for cooperation based on information and communication technologies (ICT). Trust is transmitted on the persistence and size of industrial enterprises, the economic order, the corporate culture, on positions and visions, the elasticity, the speed of the activity, the creation of alliances, the ability to create the networked organizations, on the strategy, the management, the effectivity of activities and, consequently, the competitiveness (Karczewski 2003, p. 222). Trust molds markets. In states with higher general trust (e.g., the United States of America), there emerged permanent national and transnational corporations and a lot of small- and medium-sized companies. In states with smaller trust there occurred and still are occurring small- and medium-sized companies. Moreover, there arises an essential dependence (the coefficient of correlation 0.531) between the globalization policy of a specific country and the level of general trust (Uslaner 2003). The reconsideration of determinants of general trust has consequently the essential denotation. We ought to remember that the general inclination to trust is the first stage in the process of trust building (Lipka 2005, p. 286), because it constitutes the foundations and obstacles for the whole process. It needs to be signalized that the influence of each determinant on the inclination to the total trust is dissimilar.

Literature on the subject outlined two approaches to the sources of general trust: the individualistic theory and the social theory connected with each variable (Table 1.3). The importance of these theories in the formation of general trust is continually quickly discussed in the literature of the subject. These theories are not preclusive of themselves, but rather complementary. They analyze social trust from different points of view: individuals, the local community, the country, and different stages in our lifetimes. There occurs the essential question about the logic of reasons and effects in the analysis of the social trust. How does one judge whether trust induces people to membership in voluntary associations or whether participation in them generates a trusting attitude to others? This is a pending problem. However, the true statement is that we can only consider the spiral of trust that is the mutual strengthening of the factors determining trust and trust itself.

According to Delhey and Newton (2002, p. 9) we can distinguish six theories of trust (see Table 1.3). **Personality and origins theory** concentrates on developing trust from childhood experience and cultural roots. General trust in the lifetime of a mature man changes slightly because of experiences, and if it occurs, it is caused by traumatic experience. It results from formed optimism and faith in cooperation and convictions; those individuals are able to dissolve conflicts and to live in the society satisfying the

TABLE 1.3
General Trust Theories and Factors According to Delhey and Newton

	Theory of Trust	Factors Describing the Theory
INDIVIDUAL PERSPECTIVE	1. Personality and origins theory	Optimism, life control, hesitation connivance, propensity to risk, localization of control sense (intra/extragovernability), extraversion/intraversion, generation affiliation, ethnic origins
	2. Success and well-being theory	Qualifying profile, education, income, social status, life satisfaction, job satisfaction
SOCIETAL PERSPECTIVE	3. Voluntary organization theory	Membership of voluntary associations
	4. Social network theory	Networks of friends, neighbors, groups of mutual interest
	5. Community theory	City/town/village size, community safety, satisfaction with the community, social coherency
	6. Societal theory	Public safety, satisfaction with democratic institutions, political freedom, social conflicts

Source: Adapted from Delhey, J., and Newton, K. 2002. *Who trusts? The origin of social trust in seven nations.* FS III 02–402. Research Unit "Social Structure and Social Reporting," Berlin, p. 9.

whole public. Optimism is measured by relation of the following type of the statement "the fate of an ordinary man worsens," "it is not honest bringing on the world a child." A sense of internal control over the environment is rated using trust level in science. The stronger the sense of personal control, the higher the ability to trust. The inclination to take risk is positively correlated with given trust. Trust can be found as a strategy for dealing with uncertainty in conditions of limited amount of information. An aversion to risk, low aspirations, passivism, pessimism, attachment to the past, and endeavor to maintain the status quo are orientations linked to distrust.

Personality and origins theory is depicted by Uslaner (2002). He affirms that general trust is a result of learning from parents and the nearest environment. To a certain level, trust reflects practical experiences, but first of all, honesty of the man and his morals. Behind such trust stands the conviction that most people agree on similar moral principles. Putnam (2000, p. 135–136) uses even honesty as a substitute for the trust measure. Moral trust is grounded in the conviction that the world is a good place to live, people are generally well motivated and belong to the same moral community, although they can have other origins and religious backgrounds (Uslaner 2003, p. 3). It is an effect of early learning from parents. Uslaner (2003, p. 6) even takes the view that later experiences have no influence on the general inclination to trust, except for small facts, such as higher education.

It is pertinent to trust phenomenon that it has strong cultural roots. According to Fukuyama (1995), it is a mechanism based on the assumption that other members of the certain community are characterized by honest and cooperative behavior on mutually shared rules. General trust is relatively permanent in space and time. Trust level in different countries through years is strongly correlated. For example, the trust level in countries between the years 1981 and 1990 is strongly correlated (0.81).

TABLE 1.4
Ethnic Groups and Their Levels of Trust

Group	Countries of Origin	Level of Trust for Certain Group	Countries	Level of Trust in Countries
Nordic	Sweden, Norway, Finland, Denmark	**0.592**	Sweden, Norway, Finland, Denmark	**0.58**
German	West Germany, Austria	**0.465**	West Germany	**0.42**
British	England, Scotland, Canada, Australia	**0.535**	Great Britain	**0.44**
Irish	Ireland	**0.488**	Ireland	**0.47**
French	France, Canada (French part)	**0.478**	France	**0.23**
Italian	Italy	**0.393**	Italy	**0.35**
Spanish	Spain	**0.301**	Spain, Latin America	**0.213**
Eastern Europe	Czech Republic, Hungary, Lithuania, Romania, Yugoslavia	**0.47**	Armenia, Czech Republic, Hungary, Lithuania, Poland, Romania, Serbia, Slovakia, Slovenia, Ukraine, Croatia	**0.257**
Russian	Russia	**0.479**	Not applicable	**0.24**
African	African countries	**0.183**	Afro-American countries	0.167

Source: Uslaner, E. 2004. *Where you stand depends upon where your grandparents sat: The inhabitability of generalized trust*. Working Paper. University of Maryland: College Park, p. 22. With permission.

In paneling study 1972-74-76 American National Election Study (ANES), questions concerning general trust show a large stability—75% of respondents' answers remained the same in time (Uslaner 2004, p. 11).

Trust is not only stable in time but also between generations. This denotes that people from cultures with the highest trust (e.g., Swedes, Finns, Norwegians, and Danes) will be most trusting aside from the place they live. "Patrimonial trust" from a family has a primacy over later experiences of an individual. According to this approach the change of residence country, settlement in the community demonstrative the higher trust should not significantly change already formed the general trust of the individual. For trust generally speaking the higher meaning have ancestry than the neighborhood of certain individual. Trust reflects the approach of a particular individual and it is an effect of the persistence and the mobility of a specified culture. Putnam (2000, p. 294) noticed that social capital is greater in states of the United States that have more Nordic immigrants. This indicates that one's origins shape trust in at least two ways. Firstly, being a Scandinavian makes people more trusting. Secondly, the life between Scandinavians indissolubly prompts to higher trust, even if the entity has other cultural roots. These regularities are illustrated in Table 1.4, which presents data from research between 1972 and 1998 from the General Social Survey in which trust levels among people of different nationalities living in the United States were compared with levels of trust in the countries of origin. The most trusting groups are Scandinavians, Britishers, and Germans. People who are from countries with high trust still demonstrate that higher trust in the United States. Immigrants from Eastern Europe, France, and Russia trust more than their countrymen in their countries of origin. This can be explained better in the case of immigrants' education, an influence of different institutions on the inclination to trust. We can deduce that despite an inborn inclination to general trust, we cannot ignore the environment in which a particular individual exists.

Success and well-being theory emphasizes the individual experiences of individuals during the lifetime and widely understood capital resources (Sztompka 2005, p. 320)—education, property, relationships and acquaintances, family support, the state of health. Generalized trust is the result of the experience of good treatment. People with high social status, high job satisfaction, and high income who are treated by others with respect are more willing to trust than those who experience the sacrifices of hardship, unemployment, discrimination, exploitation, and social exclusion. Trust entails a risk of losses that cannot be accepted by the poor. Each capital resource belonging to an individual determines the specific insurance against life threats, also against these that are connected with the badly located, disappointed trust—hence, the greater frankness and readiness to take risk, which carries the decision about trust. Lower income is also repeatedly a reason for the diminution of everyday optimism and a sense of control, elements that, according to the success and well-being theory, are factors determining the inclination to trust. On trust influence also relatively new traumatic experiences of individuals (e.g., divorce, separation, disease, poor financial situation, and other different kinds of failures). The influence of traumatic factors on the fall of trust refers not only to events experienced directly by an entity but also by the nearest environment of this entity. Statistically essential to the explanation of the inclination to trust (the coefficient of correlation amounts 0.81) is the number of years of education (Sztaudynger 2003, p. 236).

Very interesting are the experiences of Professor W.M. Grudzewski referring to the subject of the theory of the success. The professor was asked one day by his students: "Sir, Professor, how does one succeed in scientific research, designing, managing, and didactics?" The professor ascertained:

1. First, you must work, work to the very end.
2. When a glimmer of success appears in the tunnel, you cannot rest on a laurel or two laurels. You need to still insistently work very hard.
3. When you make greater successes, then you need to have a false friend or false institution which "will put the man to the hog," because "the hog put to the man" in the suitable period is a friend of the man; often it is the impulse to the further activity.
4. When you reach full success, then it is necessary to ask all people around for forgiveness, because if your success will not be forgiven to you, then no one will acknowledge it.

Community theory focuses on features of the local community in which an individual exists rather than on social ties existing there. House and Wolf (1978) argue in their own research the affirmative inverse dependence between level of general trust and size of a city, town, or village. Residents of small cities and villages are more altruistic, sincere, and willing to trust (Putnam 2000, p. 205). Other elements taken into account in this theory are satisfaction from living within a community and a sense of safety at night. The affirmative influence on trust development has the homogeneity of the particular community, according to the liking these which are similar to us, that covers family members, the same social group, ethnic, races. Societies in which certain discriminated minority groups live are characterized by a lower inclination to trust. Countries with large ethnic differentiation and large polarization also demonstrate smaller trust levels (Alesina and La Ferrara 2000, p. 1). In community theory, the following question often appears: To what extent is the general trust level of a particular individual shaped by the general trust level of the society in which he or she lives? In order to answer that question, research was conducted to compare immigrants' willingness to trust in the country of their origin and in the country of their immigration. The time of habitation in a particular community has an influence on trust. Trusting relationships are also developed by material culture included in architecture circumjacent individual (Borowik and Borowik, 2003, p.168). The relation to the environment is influenced by the open space of meadows, forests, and lakes, by whether the gates of the house or school are hospitable for the other man whose guard the door phones, candid cameras, high fences, Venetian blinds, or bars in windows.

The approach "from below" is somehow an alternative for the idea "from the top," namely **societal theory.** It explores the role of all public institutions in creating the social trust. Countries with high social cohesion demonstrate greater trust than countries with large social polarization measured with the differentiation of earnings in the society, class partition, and ethnic homogeneity. Societies with independent courts, efficient institutions of the authority inspection, are more trusting. In societies with efficiently working institutions, an individual feels protected by law in cases of extremely noncooperative behaviors.

In the illumination of the disposition to trust frequently takes under the attention demographic variables such as age and sex. It is possible that social trust proceeds according to U Yerkes-Dowdson curve. Youths are less trustful (Uslaner 2002). Patterson (1999, p. 173) discovered that women trust less than men in the United States (similar research confirming this regularity was done in Germany). This can be a result of the discrimination of women, their protecting role of children, and other oddities of the female sex.

In Poland, there is no systematic research on general trust. Data from the Polish Social Opinion Poll (Borowik and Borowik, 2003, p. 168) brought to daylight the information that in 1994, 90% of Poles responding voiced an opinion that "there is never too much caution in relationships with other people." Meanwhile, only 61.4% of Americans agreed with this answer. With the ascertainment "we can trust most people" only 8.3% of the Poles agreed. Among young people it was found that there is an even higher distrust (till 66% agreed with the ascertainment that "we cannot trust anybody"). From 2001 to 2003, there was done research embracing approximately

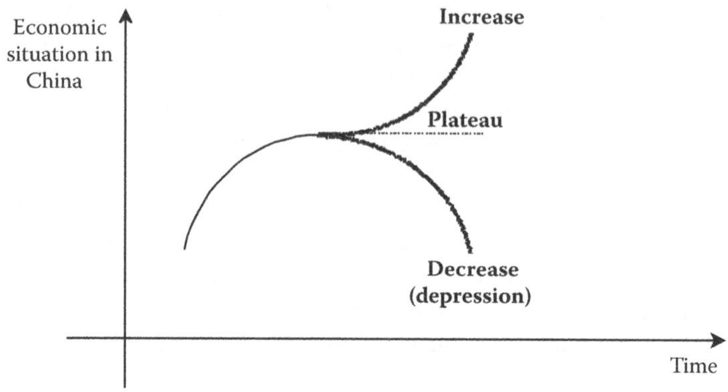

FIGURE 1.13 Forecast of the development of the Chinese economy depending on Chinese culture.

83 students and managers from small- and medium-sized from Cracow and nearby regions (Bjerke 2004, p. 302–306). Respondents were asked to express their opinions concerning trust in others in comparison to five other cultures: American (the United States), Arabic, Chinese, Japanese, and Scandinavian. The average trust is nearer to the level of the Japanese culture than American and Scandinavian. Japanese people as well as Arabs belong to cultures in which a norm is a general mistrust behavior of "foreigners." Japanese people cooperate compactly with themselves; between themselves they can hold very close relationships and are loyal to their firms. But *gaijin* ("the foreigner") will not be fully accepted by them. The Arabs' approach boils down to mixed feelings about the work done with members of other clans or tribes. Chinese peoples do not place trust in anybody from outside the (extended) Chinese family. If they come from different groups, then they have usually large problems with cooperation. This fact can have a strong negative implication in the future economic growth in China (Figure 1.13). It can stop progress on a certain level (the economy will plateau). The worst-case scenario is that it will cause a depression (a sharp decline in the Chinese economy). However, a possible third forecast is that there will appear some external factors that will help to grow this economy. An example of this factor can be some changes driven by young Chinese, educated all over the world. This group can influence a change in culture so that Chinese would start to trust others outside their family. What is more, Americans and Scandinavians belong to cultures with the highest levels of trust (Bjerke 2004, p. 252).

We can raise a question: What features of a culture create favorable conditions for trusting relationships? Coming to the point, we could analyze the determinants of the Swedish culture in comparison with other systems according to seven dimensions proposed by Hampden-Turner and Trompenaars (2006, p. 248; Table 1.5). From this research, it is obvious that countries with the highest general trust (such as the United States and Sweden) are in six categories similar to themselves (Hampden-Turner and Trompenaars, 2006, p. 248). Both cultures are characterized by the considerable universalism, an inclination to analysis and individualism, and the idea of morals as a voluntary contract that qualifies rules of the conduct. Universalism emphasizes taking into consideration the general rules in problem situations that make human activities more foreseeable. It concentrates more on rules than on human relationships.

Individualism favors the personal responsibility for appointive targets and reduces the occurrence "of hiding itself behind somebody else's back." Collectivism fosters cooperation, exchange of information, and creation of ties. However, Swedes present the individualism planned, followed by conviction that what conveys the freedom should be used for the social good. More than representatives of other cultures, they want to qualify their own identities on the grounds of their own contribution to work and relationships with friends. The rejection of hierarchy and the equality promote trust in compliance with the theory of "liking similar to us." The difference appears in external controllability of Swedes stemming from the strong dependence of their economy on the export and limited resources of labor.

Cultural variety determines our disposition to trust and portions out trusting culture and culture of distrust (cynicism). Every dimension of a culture and the form it takes determines the source of trust. According to the classification of Hofstede's cultural dimensions (1984), the small/large power distance, the collectivism/individualism, the femininity/masculinity, the weak/strong avoidance of the uncertainty—across these factors we can qualify sources from which trust is generated (Figure 1.14): the calculation or the familiarity.

The indication on the masculinity and the femininity of the culture assumes that there exists certain regularity in the social roles credited to both sexes and way of behaviors. Masculinity is identified with the endeavor to success, professional and material successes, in opposition to the feminine culture, which values human relationships, quality of life, and care. Male culture is based more on trust generated more from the economic calculation of advantages and losses than from empathy, care, and common values, as in the case of the feminine culture. The dimension of individualism gives preference to the good of the individual over the public welfare. It is based on the perception of the entity as the independent human being, individually making decisions about its own fate and carrying all consequences of its own actions. Relationships with other partners, also independent units, spring from the rule of contracts and exchanges. Collective behaviors appear wherever the priority is the good of the group. The fondness for our own group is simultaneously accompanied by antipathy to other groups perceived as strangers. Trust derives from the familiarity of the group's members. The distance to the authority refers to the range of acceptances and expectations against the unequal distribution of the authority in the organization and the society. The small power distance, and at the same time the greater equality, create the conditions necessary for trust founded on mutual relations. Trust in the large power distance culture is based on formal authority. Avoiding uncertainty defines the level of tolerance of the uncertainty and the amphiboly of new situations. The avoidance of uncertainty is interlinked with the human need of the order. In this instance of strong uncertainty avoidance there appears support for ideologies and institutions assuring the trimness of the reality, making life predictable. In a culture with weak uncertainty avoidance, people are more likely to run risks. Dimensions of the culture in Poland according to Hofstede (1984) are presented in Graph. 1.2. This graph shows that Poles achieved comparatively higher results with respect to uncertainty avoidance (106, while the Hofstede max is 112) as well as power distance (72, while the Hofstede max is 104). The empirical results of Polish managers in the field of masculinity accounts for 62, and the Hofstede

TABLE 1.5
Seven Dimensions of Culture in Different Countries According to Hampden-Turner and Trompenaars (2006)

Universalism	Particularism
Rules, codes of law, and generalization  **The United States** **Sweden** **Germany**	Exceptions, special conditions, unique relations  **France** **Japan** **Germany**
Individualism	**Collectivism**
Personal freedom, human rights competitiveness  **The United States** **Sweden**	Social responsibility, harmonious relations, collaboration  **France** **Japan** **Germany**
Fragmentation	**Entireness**
Atomization, analysis, objectivism  **The United States** **Sweden**	Holism, polishing, synthesis, relation ability 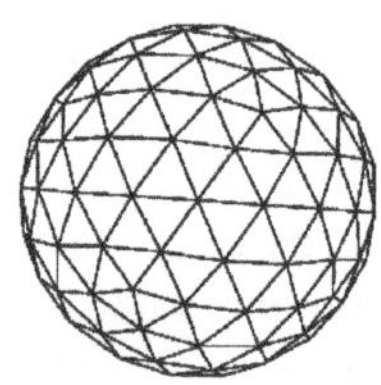 **France** **Japan** **Germany**
Internal Controllability	**External Controllability**
Conscience and internal beliefs  **The United States** **France** **Germany**	Patterns and factors influencing outside 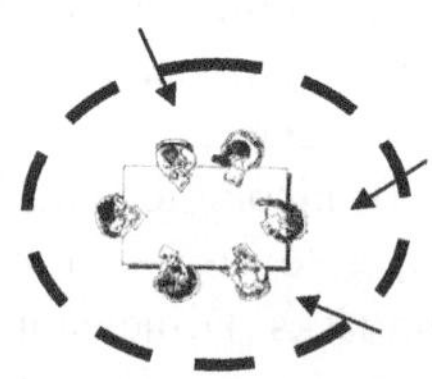 **Sweden** **Japan**

(continued)

TABLE 1.5 (CONTINUED)
Seven Dimensions of Culture in Different Countries According to Hampden-Turner and Trompenaars (2006)

Gained Status		Ascriptions Status	
What we did, our results **The United States** **Sweden** **Germany**		Who we are, our potential, connections **France** **Japan**	
Equality		**Hierarchy**	
The United States **Sweden** **Germany**		**France** **Japan**	
Sequentiality		**Synchronousness**	
Keeping with scheduled course **The United States** **Sweden**	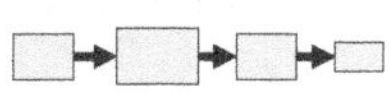	Tasks coordinating **Japan** **France**	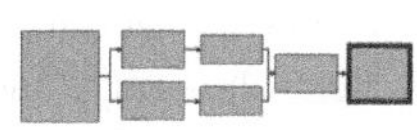

Source: Adapted from Hampden-Turner, C., and Trompenaars, A. 2006. *Siedem kultur kapitalizmu: USA, Japonia, Niemcy, Francja, Wielka Brytania, Szwecja, Holandia.* Kraków: Oficyna Ekonomiczna. Oddział Polskich Wydawnictw Profesjonalnych.

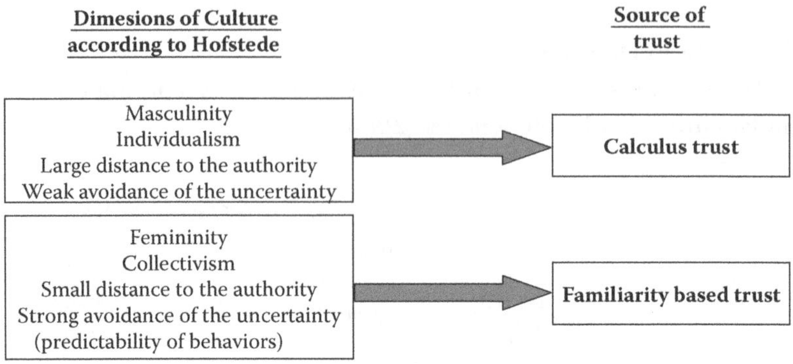

FIGURE 1.14 Trust dimensions of culture according to Hofstede (1984) and sources of trust.

max is 95. The lowest results Poles gained is in individualism (56, while the Hofstede max is 91). However, we can look at the details and compare our country with other groups (clusters) of countries. Japanese culture is characterized with higher masculinity level (95) than Polish culture (only 62). The most individualistic is the Anglo-Saxon cluster (83).

Generalized trust is so to say a feature universally prevalent, but with different intensity. This is a specific continuum of trust from the mistrust across limited and complete trust. It is a constant relatively feature, what does not make it invariable. We can use different techniques for building generalized and particular trust. There would not be, any rules how to develop trust, training, if we attended, that trust management is impossible.

What advice for trust management is revealed from the generalized trust theory?

1. The aptitude to trust in a concrete situation is determined by a certain relatively constant predisposition to trust.
2. Do not forget about the personality traits in trusting relationships.
3. Sex, age, and education impact trust.
4. The cultural variety of an individual determines generalized trust.
5. One's religion and beliefs system can be an obstacle in developing trust.

Economic subjects work in the existing system of the trust culture that is characteristic for the given national culture.

The above guidelines are followed by a lately intensely developed discipline called **diversity management** because trust apprehended as the element of cultural foundations. Diversity management first refers to building consciousness and acceptance of sociodemographic differences in an organization and contains activities that facilitate benefits from the differences in a way that is profitable for the organization (Roberson 2004, Woźniakowski 2005). Typical activities that can be realized within the framework of programs of the diversity management in order to build trust can be as follows:

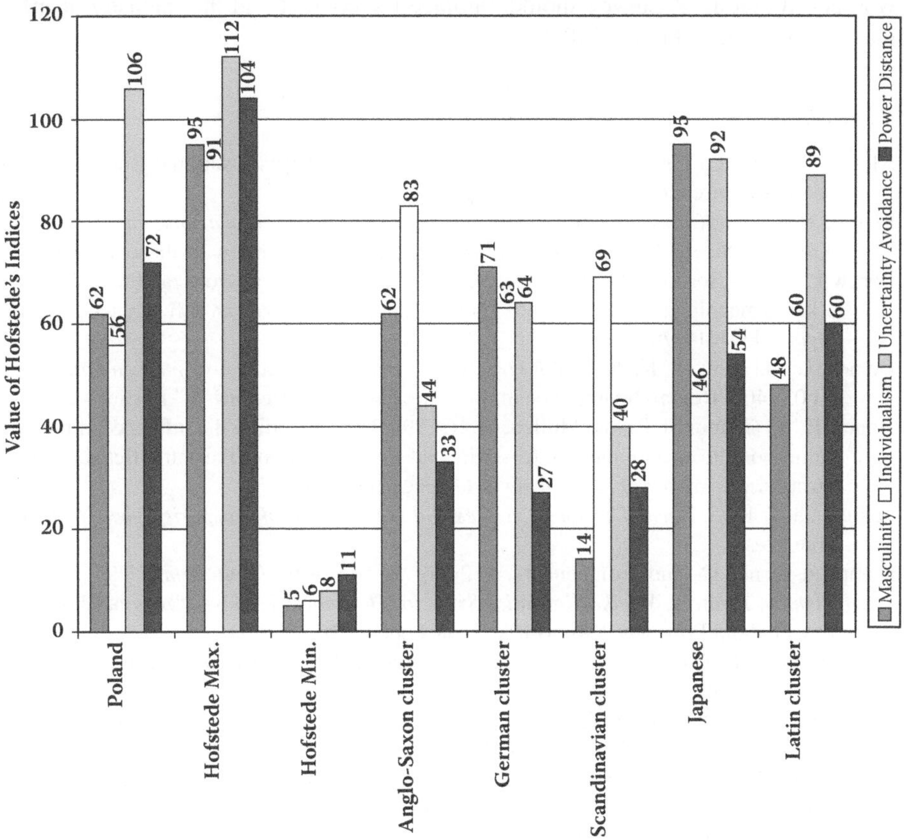

GRAPH 1.2 Culture dimensions of Polish managers according to Hofstede's indices. *Source:* Adapted from Nasierowski, W., and Mikuła, B. 1998. Culture dimensions of Polish managers: Hofstede's indices. *Organization Studies* 19/3, p. 495–509.

- Building diversity consciousness among managers and workers of a firm so as to change the attitude of workers and diminish the risk of the appearance of behaviors that can be negatively understood by groups representing minorities in the firm.
- The propagation of programs of diversity management by instruction, seminars, coaching, mentoring, building of teams, building contacts, and exchange of information.
- Adapting the procedures of a firm within the range of the recruitment and selections, the estimation of workers.
- Workers' indoctrination to what the differentiations are within their teams and how these differences can be used for the common good.

We can never assume first of all that trust exists in an organization by nature. Careful diagnosis of this matter should be included in any systematic research on workers' opinions. A survey should embrace the content and the intensity of information exchange (Marcisz 2003).

REFERENCES

Alesina, A., and La Ferrara, E. 2000. The determinants of trust. National Bureau of Economic Research, Cambridge, Working Paper 7621.

Bjerke, B. 2004. *Kultura a style przywództwa, zarządzanie w warunkach globalizacji.* Kraków: Oficyna Ekonomiczna. Oddział Polskich Wydawnictw Profesjonalnych.

Borowik, R., and Borowik, B. 2003. Kapitał-społeczno-kulturowy—rola zaufania jako czynnika wzmacniającego efektywność polityczną, gospodarczą i cywilizacyjną. *Prakseologia*, 143, 161–178.

Delhey, J., and Newton, K. 2002. *Who trusts? The origin of social trust in seven nations.* FS III 02–402. Research Unit "Social Structure and Social Reporting," Berlin.

Doney, P. M., Cannon, J. P., and Mullen, M. R. 1998. Understanding the influence of national culture on the development of trust, Special Topic Forum on Trust and Between Organizations. *Academy of Management Review*, 23(3), 1–17.

Fukuyama, F. 1995. *Trust: The social virtues and the creation of prosperity*, New York: The Free Press.

Hampden-Turner, C., and Trompenaars, A. 2006. *Siedem kultur kapitalizmu: USA, Japonia, Niemcy, Francja, Wielka Brytania, Szwecja, Holandia.* Kraków: Oficyna Ekonomiczna. Oddział Polskich Wydawnictw Profesjonalnych.

Hofstede, G. 1984. *Culture's consequences: International differences in work-related values.* London: Sage.

House, J., and Wolf, S. 1978. Effects of urban residence on interpersonal trust and helping behaviour. *Journal of Personality and Social Psychology*, 36, 1029–43.

Karczewski, L. 2003. Założenia kulturowe- zaufanie-zarządzanie zasobami ludzkimi w Stanach Zjednoczonych i w Japonii. *Prakseologia*, 143, 217–229.

Knack, S., and Keefer, P. 1997. Does social capital have an economic payoff? A cross-country investigation. *Quarterly Journal of Economics,* 112, 1251–88.

Kożusznik, B. 2005. Wpływ a zaufanie. In *Wpływ społeczny w organizacji.* Warszawa: Polskie Wydawnictwo Ekonomiczne.

Lipka, A. 2005. RóŻnorodność kulturowa a budowanie zaufania w organizacji. In *Praca i zarządzanie kapitałem ludzkim w perspektywie europejskiej*, A. Pocztowski (ed.). Warszawa: Polskie Wydawnictwo Ekonomiczne, 279–288.

Marcisz, A. 2003. Istota zaufania i jego pomiar. *Prakseologia*, 143, 153–160.

Nasierowski, W., and Mikuła, B. 1998. Culture dimensions of Polish managers: Hofstede's indices. *Organization Studies* 19/3, 495–509.

Patterson, O. 1999. Liberty against democratic state: On historical and contemporary sources of American distrust. In *Democracy and Trust*, M. E. Warren (ed.). Cambridge: Cambridge University Press.

Putnam, R. 2000. *Bowling alone: The collapse and revival of American community.* New York: Simon & Schuster.

Rotter, J. B. 1967. A new scale for the measurement of interpersonal trust. *Journal of Personality,* 35(4), 651–665.

Stolle, D. 2001. Clubs and congregations: The benefits of joining an association. In *Trust in society,* K. S. Cook (ed.), 202–244. New York: Russell Sage Foundation.

Sztaudynger, J. 2003. Zaufanie i kapitał społeczny a wzrost gospodarczy. Wyniki badań ekonometrycznych. *Prakseologia,* 143, 231–245.

Sztompka, P. 2005. Kultura zaufania. In *Socjologia, Analiza społeczeństwa*, Warszawa: Znak.

Trompenaars, A., and Hampden-Turner, C. 2005. *Zarządzanie personelem w organizacjach zróżnicowanych kulturowo*, Kraków: Oficyna Ekonomiczna. Oddział Polskich Wydawnictw Profesjonalnych..
Uslaner, E. 2003. *Trust in the knowledge society.* Conference on Social Capital, Cabinet of Government of Japan.
Uslaner, E. 2004. *Where you stand depends upon where your grandparents sat: The inhabitability of generalized trust.* Working Paper. University of Maryland: College Park.
Uslaner, E. M. 2002. *The moral foundations of trust.* New York: Cambridge University Press.
Woźniakowski, A. 2005. Doświadczenia firm amerykańskich w zarządzaniu róŻnorodnością i uwagi na temat możliwości wykorzystania ich w warunkach polskich. In *Praca i zarządzanie kapitałem ludzkim w perspektywie europejskiej*, A. Pocztowski (ed.). Warszawa: Polskie Wydawnictwo Ekonomiczne.

1.5 TRUST VERSUS CONTROL

Trust and control are welded. A never-ending discussion debates whether they are substitutive, supplemental, complementary, or antagonistic (one eliminates the other) to each other. Solvations are different for concrete cases, thus we can adapt here the situational approach. To strengthen the emphasis of these diverse dependences encapsulated, Möllering (2005) defined the trust-control relation as the duality. Nevertheless, it is essential to understand the relation between trust and control in order to work out an effective plan for developing a trusting relationship. Trust also has to perform functions that in traditional organizations are fulfilled within control, such as assuring assignment accomplishment in compliance with intentions and correcting faultiness by means of feedback about the course of processes. Generally, trust and control help to reduce perceived risk to some acceptable level, making the players' behavior more foreseeable. Skinner and Spira (2003, p. 29) acknowledge trust as the strategy of risk management and uncertainty management.

In the section "Theoretical Approaches toward Trust," the essence of trust was introduced. There comes into being the question for the essence of control. **Con trol** is generally defined as the process of regulation and monitoring necessary for accomplishing previously established targets, assuring that real activities will conform with planned ones (Stoner et al. 2001). "Well-placed balances on the border of two worlds: the past and the future trying not to entangle too strongly into their structures, not to waste equilibrium" (Kuc 2001, p. 14). Green and Welsh (1988) delineated control as the cybernetic, regulative process that directs and limits interactive activities to certain standards and aims. In this context, control assures the realization of assignments according to plans, across the checking their realization and stimulating the corrective activities when some perceived deviations (in the feedback) appear.

Thereafter, **organizational control** is recognized as any mechanism that managers use to ensure that subunits within an organization act in a coordinated and cooperative way in order to achieve the organization's targets by the best possible allocation of all resources given the conditions (Lebas and Weigenstein, 1986). Managers tailor their organizational control tools (applications) directly to their subordinates' activities and ongoing processes.

The **function of control** should be performed according to the following rules (Kuc 2001, p. 116):

- Controlment is a systematic effort undertaken for the qualification of patterns (norms, standards) of activities, comparing them with reached performance and starting off, if necessary, corrective activities.
- The reason for existence of control is making both measurements of results as well as deviations from patterns.
- The efficiency of control depends on the speeds of the reaction for defections and not only on fact of their ascertaining.
- Final control (control of results) gives place to anticipatory control overreaching the event (results of events).
- Selective control is the most effective form of control in stopping trouble for the success of the undertaking.

There subsist many ideas of control (presented for instance by H. Emerson, K. Adamiecki, A.W. Gouldner, and many others scholars). We ought always to remember first and foremost that control is a tool to compare real performance with organizational plans and to find the size of deviations. As a matter of fact, control is predominantly analyzed as the phase (the inseparable element) of the organizational action cycle (Kuc 2001, p. 47).

Managerial perspective distinguishes three focal types of control:

1. **Input control**—for example, training socialization that helps to guide the selection and preparation of applied resources, material as well as human ones (Arvey 1979, Van Maanen and Schein 1979, Wanous 1980)
2. **Process control**—utilized to ensure that subordinates implement all necessary production methods
3. **Output control**—to inspect that the output employees produce meets the performance standards (Ouchi 1977, 1979, Mintzberg 1979).

Control mechanisms can be grounded on processes, results, social elements, and self-control. The **characteristics of control mechanisms** according to that distinction are as follows (Bradley 2003):

1. **Process/behavior:**
   - **a.** Development of rules and procedures of conduct, which help to attain the purpose essential for the organization
   - **b.** Monitoring and estimation of activities in every unit, which assures that all required procedures are complied with
2. **Result/outcome:**
   - a. Establishment of common standards and measurable targets
   - b. Monitoring and estimation of whether the gained results for every activity help attain the intended aims and standards realizations
3. **Social elements (adherence):**

a. Limitation of human behaviors by internationalization processes of values and group norms
b. Cooperation of entities to attain goals because they take part in internationalization (processes of internationalization demand cooperation and mutual support).

4. **Self-control:** Entities agree on their own aims and are internally motivated to realize assignments and reach their aims. They control themselves and rate their own activities. It is the opinion very consistent with the theories of Maslov (1943) and Herzberg (1968) of the needs of self-actualization among workers. Employees representing the highest qualifications realize attached to their assignments partially and individually at the same time. Furthermore, assignments are formulated the way that they encourage to qualifying independently the methods, manners, and resources. This stimulates self-improvement and development of the knowledge and skills necessary to use information.

As far as control mechanisms based on processes, results, and social elements are objects of consideration in cases concerning cooperation, insomuch that self-control refers more to workers' motivation to the independent work and expresses the large trust into their skills. The effect of the use of control mechanisms has to be an increase in efficiency and effectiveness of team activities, organizations, and so forth.

There are two basic approaches to control: **external** (founded on measurement and processes) and **internal** (based on social values) (Eisenhardt 1985). The first conception is also called **formal control** (Jaworski 1988), and it originates from formal rules, procedures, and politicians. The second is also called **informal control** (Jaworski 1988) and is derived from organizational norms, values, culture, and internalization of aims that encourage definite behaviors and results. Formal control is performed by the observation of deportment or by the measurement of deportment's results. The observation of behavior has to assure a proper realization of processes, while the measurement of results (control of results) relies only on the suitable and reliable estimation of the work results. Every one of these forms of control has other spheres of application.

Eisenhardt (1985) suggests that picking certain suitable control mechanisms depends on two factors: task programmability and result measurability (Figure 1.15). Task programmability covers the level to which it is possible to understand the operations. In other words, low programmability means that the assignment is composite, not very transparent. The programmability of assignments is low when activities are not routine, a result of adaptation to continuously changing circumstances on the market and varying needs of consumers. Outcome measurability reflects the possibility of the objective and precise measurement of results of the activity.

With such definite conditions, control of results (Ouchi 1977, 1978) is suitable for activities, with high outcome measurability and low programmability of assignments—in other words, in situations where our knowledge about operations is large. The settlement of measurable measures, and their values and deviations diminish

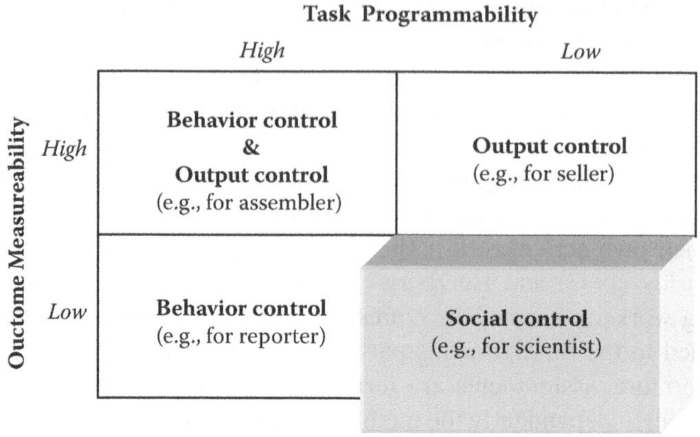

FIGURE 1.15 The control mechanism depending on conditions. *Source:* Das, T. K., and B.-S. Teng. 2001. Trust, control, and risk in strategic alliances: An integrated framework. *Organization Studies*, 22/2, p. 260. With permission.

performance risk. The most widespread control mechanisms of results are planning and budgeting.

Behaviors' control (Ouchi 1977, 1978) is also useful in other circumstances—at the low measurability of conducts and the high programmability of task, that signifies the large knowledge about processes. Behavior control is also called **process control** because it concentrates on processes that transform definite deportments into desirable results rather than on final performance. Often, because of that, the measurement of results in a given situation is not possible or is imprecise and processes are weakly diagnosed. When the nature of cooperation is not complicated and there exist standards for products and services usually no one controls the act of the cooperation. Such control reduces the rational risk, which is connected with the probability of crafty and undesirable activities whose results are very hard to measure. Examples of widely used control mechanisms are policies, procedures, reporting structures, training, and staffing. Procedures and policies can describe what and how something has to be made. Reporting structures qualify the parts played by every subject in the system, assuring supervision and monitoring of activities. Staffing and training regulate and standardize behaviors to the given level of the circumstances. With suitable procedures of recruitment, selection, and staffing, it is possible to select workers with conforming, required behaviors. We can also convince workers to perform certain behaviors through training.

When both measurabilities and the programmability of assignments are low, there enters social control (Jaworski 1988), which lets us participate in the settlement and the realization of aims. Social control targets the reduction of the divergence in preferences of members of the organization regarding targets across the universally recognized culture and value system (Kirsch 1996). The difference between formal and social control (with behavior and outcomes) originates from the fact that neither the deportment nor results are compactly defined at the beginning. There are no rigidly settled borders of activities. The process of the settlement of aims is decentralized and develops in time; that is why social control is suitable for ambiguous situations, when the usage of outcome control and processes control are limited. Through the socialization process, members become more devoted to their organization and share opinions that impact the behaviors in the work process, reducing rational risk and the risk of failure in outcomes. Mechanisms of social control include exhibiting and the spread of the culture across rituals, ceremonies, and cooperation in order to settle common norms and convictions. Social control is nearest to the notion of trust and is often also deceived with it. Mechanisms of social control are the elements in enactment of trust.

Control mechanisms have to protect from undesirable results of opportunistic activities of the second party. Graph 1.3 presents the methods of protecting in case of failure in the cooperation of small- and medium-sized enterprises from trade, services, and manufacturing sectors. In his research, Schumacher (2003) discovered that 10.4% of small- and medium-sized enterprises use no protection (Graph 1.3). The risk of opportunistic behaviors exists however in every business relation. The reason that 1 of 10 firms decides to expose itself to risk in spite of the lack of protection lies in social control mechanisms determining trust, which make the perceived risk of activities smaller. Moreover, the mechanisms of social control are not often as commonly understood as the mechanisms of control. Actually, the lack of the psychological consciousness of this fact makes social control more effective. Some firms decide to have several protections, whereas others have a smaller number at the higher level of trust.

We raise the following question: Does trust influence control framework and, if it does, how? This depends of course on the concrete situation and circumstances. In compliance with the intuitive understanding of trust in social interactions, we ought always to take into account the so-called "tribal wisdom" because actors of the social game agree on its significant role. This commonsense wisdom is as follows: "If you control me, you will not trust in me." In this context, it is also acknowledged

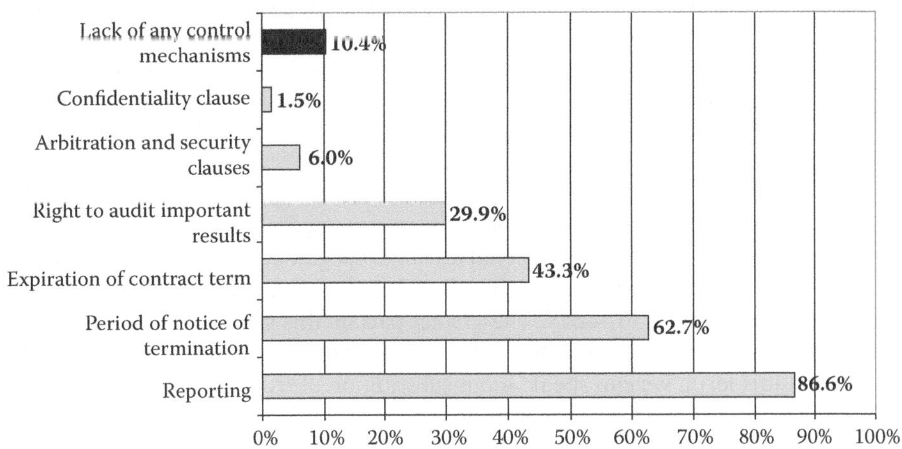

GRAPH 1.3 Safeguard methods applied in relationships with business partners from small- and medium-sized companies (n = 191). *Source:* Adapted from Schumacher, C. R. 2003. *Trust versus control in co-operative relationships of small- and medium-sized enterprises.* New Zealand Association of Economics Conference, 25-27 June 2003.

that "if you do not trust me sufficiently, you will most likely monitor or control me." Applying inflexible rules and purposes in the form of control of results and processes can lever trust, at the same time creating a mood of distrust. It is also an impeachment of the competence of organization members, their abilities and skills to perform assignments in the best way (consequently, it is seen as a negation of trust to the partner's competence [the cognitional trust]) by limitation of their autonomy in the activity and consequently the limitation of the possibility by which they can prove their own potential. Mistrust creates conviction about necessity to check frequently the second party. From this standpoint, *control* in the narrow meaning of this word is an antagonist of trust (Castelfranchi and Falcone 2000). Formal control mechanisms (concerning the results and processes) can be recognized as weak impersonal substitutes for trust. On the other hand, social control, being based on common values, aims, culture, systems of convictions, and preferences, is a bracing piece intensifying sympathy and consequently trust. In this instance, partners have a larger freedom of the conduct because they have not got any compulsory behavior. Mutual interactions (frequent meetings, communication, cultural ties, and socialization) bear on the shape of future results and build both good will trust and cognition-based trust.

Indubitably trust has also an influence on control. The control of processes and the results of control are more effective when they are accompanied by trust, which was discovered in empirical research by Vryza and Fryxell (1997). Furthermore, we begin to speak about control in situations in which we undertake making decisions about the activity; this already demands at least a minimum of trust into the successful outcome of activities. A definite level of trust is necessary for cooperation to flourish. In the instance of cooperation between new partners, certain mechanisms of control are instituted to stimulate trust and consequently to encourage this cooperation. Along with the development of this cooperation and with the repetition of the interaction, this affirmative relation between control mechanisms and trust can permute. Increasing trust between cooperative firms should actuate a decrease in the level of required control, such as contracts, systems monitoring, and so forth. Such activities are also recognized as trust intensifiers because they demonstrate the readiness to be exposed to risk, opportunistic behaviors, and faith in the good will of the second party (as illustrated in Graph 1.4).

Definite investments into assets connected with cooperation determine the essential encouragement to trusting relationships, as that suggests long-term engagement. On the stage of trust initiating, personal attributes (e.g., open fission with ideas, supporting trust) have a smaller impact because they are more difficult to estimate, less self-evident, and harder to verify. The greater part on this stage plays the resignation from formal control mechanisms. Trust entails excluding certain control mechanisms. On this level, we can speak about substitution of trust. Schumacher (2004, p. 15), on the basis of previous research, found and explored the relation between the following contributions:

x_1 - the level of trust
x_2 - the level of control.

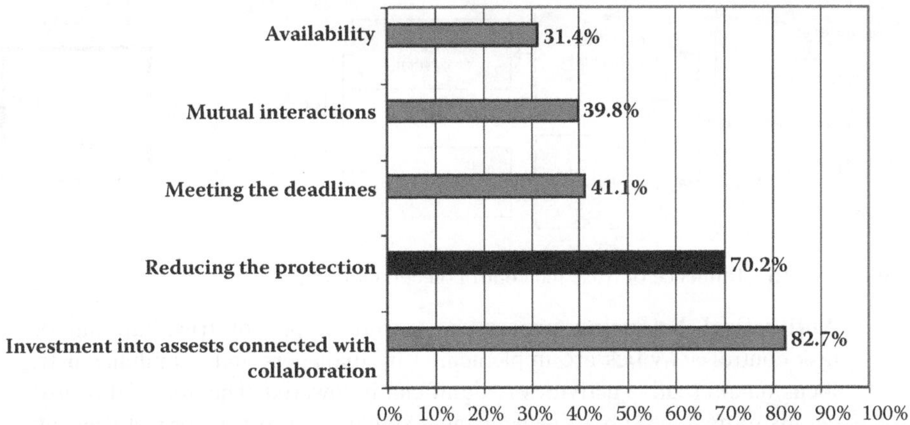

GRAPH 1.4 Trust-building behaviors in business relations between small- and medium-sized enterprises and their partners (n = 191). *Source:* Adapted from Schumacher, C. R. 2002. *The influence of trust on SME co-operation structuring and performance.* NZAE Conference 2002.

He compared them with the result (*y*), a perceived certainty of partner's activities (confidence).

This trusting relationship between the cooperating business partners (small- and medium-sized companies) can be expressed in terms of the constant of the substitution elasticity (ES) in the production function (CES):

$$y = \blacklozenge\left[\blacklozenge x_1^{-\blacklozenge} + (1-\blacklozenge)x_2^{-\blacklozenge}\right]^{-\blacklozenge/\blacklozenge} \tag{1.5.1}$$

$$ES = \frac{1}{1+\rho} \tag{1.5.2}$$

The function CES with unknown parameters ρ, η, and δ is used when we ask a question about the substitution coefficient of two elements. ES equals zero when the substitution of elements is not possible. On the basis of the above formulas, Schumacher (2004) determined ES = 0.83. The high value of this coefficient implies an ascertainment that trust and control are the parts of mechanisms generating trust in business partners and this is the relation of strong substitution and consequently low supplementary abilities in cooperation relations between firms.

The situation where only trust or only control exists is very rare. Both these elements usually coexist, influencing each other and impacting the general outcomes of the enterprise (as illustrated in Figure 1.16).

The formula for the elasticity coefficient of the production function can be applied to qualify the substitution of trust and control at attaining of economic results.

Long and Sitkin (2006) contend that there are three types of balancing processes that managers apply to combine trust-building and task-control activities:

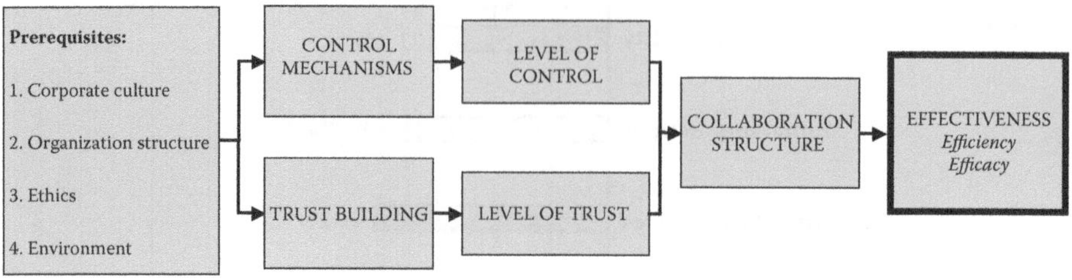

FIGURE 1.16 Influence of trust and control on collaboration.

1. **Antithetical balancing processes**—Certain types of trust-building or task-control activities are implemented by managers and, simultaneously, focus on alternative activities is significantly lowered. The aim of this process is to manage subordinates using a single form of trust or task control. The expected result is an effective implementation of trust or task control activities.
2. **Orthogonal balancing processes**—A selection of various unrelated trust-building and task-control activities is introduced. By this process, an organization manages subordinates using a combination of various unrelated trust-building and task-control activities. Moreover, there takes place constant, ongoing maintenance of multiple, sometimes contradictory, trust-building and task-control activities.
3. **Synergistic balancing processes**—A selection of different mutually reinforcing trust-building and task-control activities is implemented and managed. Synergies between multiple trust-building and task-control activities are recognized and maintained.

It is worth mentioning that expanding the control sets off for some individuals feelings of discomfort and value depreciation; this is particularly true for very creative and independent individuals. The counteraction in this instance is difficult because formal control is not free from unpleasant feelings and associations. Their sources are unpleasant facts which, when revealed, lead to criticism, thus provoking a stressful and discouraging atmosphere. Previous research shows that the occurrence of this effect was connected with evaluations that are contradictory, unascertainable, unforeseeable, or unattainable (Kuc 1999, p. 251). As far as the twentieth age experienced compressions of managerial control mostly by using scientific management rules, the twenty-first age faces changes of the controlment process. The need for control will still be strong but will be realized using new methods, such as social control, that create greater ground for the creativity of the organization, partners, and workers.

Nowadays, nobody calls into question the indispensability of control in the management process, though we can put forth the argument that well-planned resources in well-formed structures in an efficient motivation system can impose limitations on the range and intensity of control functioning (Kuc 1999, p. 251). This alternative system can stem from trust-building systems.

REFERENCES

Arvey, R. D. 1979. *Fairness in selecting employees.* Reading, MA: Addison-Wesley.

Bradley, J. A. 2003. *Institutional influences and control of software development projects: An examination of Air Force software project teams* [PhD diss.]. Florida State University, College of Business, Tallahassee, FL.

Castelfranchi, C., and Falcone, R. 2000. Trust and control: A dialectic link. *Applied Artificial Intelligence*, 14(8), 799–823.

Das, T. K., and Teng, B.-S. 2001. Trust, control, and risk in strategic alliances: An integrated framework. *Organization Studies*, 22(2), 251–283.

Eisenhardt, K. M. 1985. Control: Organizational and economic approaches. *Management Sciences*, 31, 134–149.

Green, S. G., and Welsh, M. A. 1988. Cybernetics and dependence: Reframing the control concept. *Academy of Management Review,* 13, 287–301.

Herzberg, F. 1968. One more time: How do you motivate employees? *Harvard Business Review,* February, 51–62.

Jaworski, B. J. 1988. Toward a theory of marketing control: Environmental context, control types, and consequences. *Journal of Marketing,* 52(3), 23–39.

Kirsch, L. J. 1996. The management of complex tasks in organizations: Controlling the system development process. *Organization Science*, 7, 1–21.

Kuc, B. R. 1999. *Zarządzanie doskonałe*, Warszawa: Wydawnictwo Oskar-Master of Biznes.

Kuc, B. R. *Kontrola menedŻerska. Przełamywanie barier,* Warszawa: Wydawnictwo Menedżerskie PTM.

Lebas, M., and Weigenstein, J. 1986. Management control: The roles of rules, markets and culture, *Journal of Management Studies*, 23, 259–72.

Long, C. P., and Sitkin, S. B. 2006. Trust in balance: How managers integrate trust-building and task control. In *Handbook of trust research*, R. Bachmann and A. Zaheer (eds.), 87–106. Cheltenham, UK: Edward Elgar Publishing, Ltd.

Maslov, A. H. 1943. Theory of human motivation. *Psychological Review,* 50, 370–396.

Mintzberg, H. 1979. *The structuring of organizations.* Englewood Cliffs, NJ: Prentice-Hall.

Möllering, G. 2005. The trust/control duality: An integrative perspective on positive expectations of others. *International Sociology,* 20(3), 283–305.

Ouchi, W. G. 1977. The relationship between organizational structure and organizational control. *Administrative Science Quarterly,* 22, 95–113.

Ouchi, W. G. 1978. The transmission of control through organizational hierarchy. *Academy of Management Journal,* 21(2), 173–192.

Ouchi, W. G. 1979. A conceptual framework for the design of organizational control mechanisms, *Management Science*, 25, 833–848.

Schumacher, C. R. 2002. *The influence of trust on SME co-operation structuring and performance.* NZAE Conference, 2002.

Schumacher, C. R. 2003. *Trust versus control in co-operative relationships of small- and medium-sized enterprises.* New Zealand Association of Economics Conference, June 25–27, 2003.

Schumacher, C. R. 2004. *Is trust a substitute or complement for safeguards? Evidence from Germany.* May 2004, Massey University Commerce Working Paper No. 04.06. http://papers.ssrn.com/sol3/papers.cfm?abstract_id=554483 (accessed November 11, 2006).

Skinner, D., and Spira, L. F. 2003. Trust and control: A symbiotic relationship? *Corporate Governance*, 3(4), 28–35.

Stoner, J. A., Finch, R., Freeman, E., and Gilbert, D. R. 2001. *Kierowanie*, Warszawa: Polskie Wydawnictwo Ekonomiczne.

Van Maanen, J., and E. H. Schein. 1979. Toward a theory of organizational socialization. In *Research in Organizational Behavior*, B. W. Staw and L. L. Cummings (eds.), 209–64. Greeenwich, CT: JAI Press.
Vryza, M., and Fryxell, G. E. 1997. *The interaction of trust and control mechanism in the management of successful international joint ventures*. Boston, MA: Academy of Management.
Wanous, J. P. 1980. *Organizational entry: Recruitment, selection, and socialization of newcomers*. Reading, MA: Addison-Wesley.

1.6 MODELING TRUST SYSTEMS

The notion of *system* most often is identified as the flight recorder in which are appointed "entries" and "exits" (Figure 1.17). For "entries" we attend definite influences of the system on the environment. The causal nexus exists between "the entry" and "the exit." Knowing "entry" we can foresee "exit." The system is built from the set of elements between which we can observe interrelation, where every element is connected with every other element whether impromptu, immediate or indirect.

The model constitutes a certain image, the reduction of reality. The modeling of the system stands for the representation of interesting us in the proprieties of the definite system. It assumes that the model of a given system should behave similarly to the system that is an object of the modeling (Findeisen 1985). Models of the system usually possess a considerably smaller number of entries than in the case of the real system. Depending on the destination of the model, it can accept different forms. We separate four main objectives of building of systems models (Findeisen 1985):

- **Management**—the model is to make decisions within the definite system
- **Designing**—the model targets in optimizing the structure and parameters within the projected system
- **Prognosing**—the expected behavior of the system in the future
- **Research**—the explanation of the system behavior under certain conditions.

The trust system embraces all elements directly or indirectly connected with organizational trust. *Organizational trust* refers to trust existing within the given organization. In the narrower meaning of this kind of trust, it embraces the trust of the worker to the higher management of the firm. At the same time, the model-

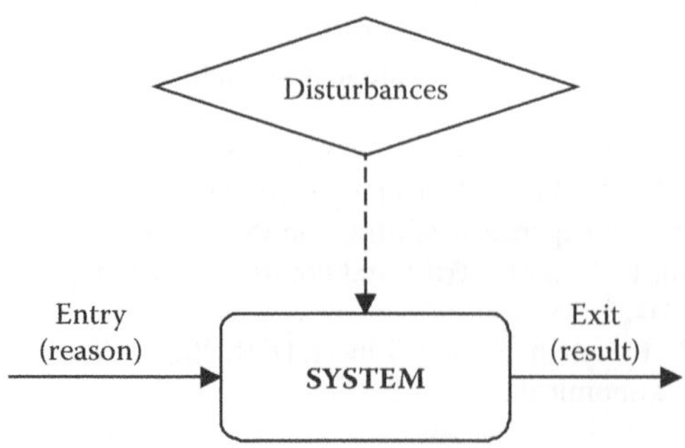

FIGURE 1.17 Model of a system.

ing of trust systems signifies showing the trust system in the concrete form. It is necessary to use a systems approach with reference to trust phenomenon because of its complexity.

Trust management demands from the enterprise an acquaintance with the level of the risk resulting from the given cooperation. Simultaneously, it raises the effectivity of such cooperation. Previous research involving small- and medium-sized average enterprises has proven this (Grudzewski, Sankowska, and Wańtuchowicz 2005; Kaufmann, Kokalj, and May-Strobel 1990).

Because of definitional problems with reference to notions of trust cause, we often describe this notion by means of models of trust systems. Several models of trust, among other things according to Lewicki and Bunker and Dyer and Chu, were introduced in the section "Theoretical Approaches toward Trust." The present section targets to present some selected, important models of trust systems.

Blomqvist Model (2004)

Blomqvist agreed that four main elements constitute trust: competences (determined by a potential), good will, behavior, and self-reference (referring to values shared in certain enterprises) (McKnight, Cummings, and Chervany 1998). The above-listed elements are derived from previous research by Blomqvist done using personal questionnaires in small- and medium-sized enterprises. All trust elements should be considered in the context of past, present, and future. It is determined by the gist of trust, which is strongly characterized by temporariness and momentariness (Figure 1.18). The modeling of trust systems is possible only in a dynamic approach in which we take into account all three time dimensions.

The trust model (Figure 1.18) consists of many dimensions, such as **competences** (for instance, technological potential, skills, know-how) and more abstract dimensions connected to **good will** within and toward enterprises. The last dimension puts an emphasis on company responsibility and the importance of positive intentions toward the environment. Contemporaneous good will of a company obliges the company to perform specific behaviors and avoid certain situations. Together with the development

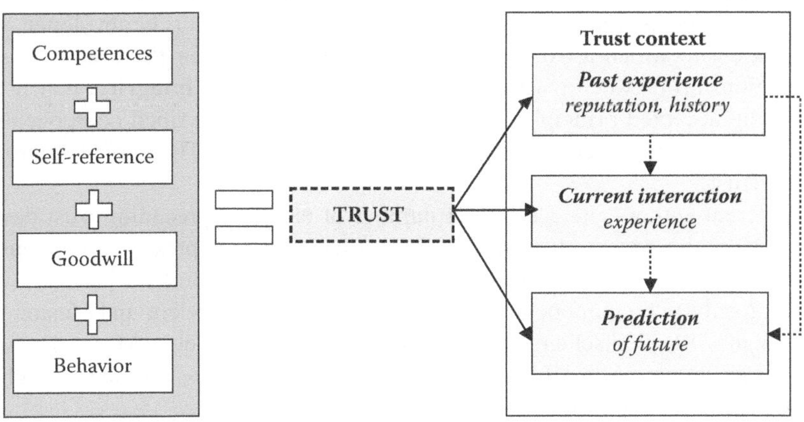

FIGURE 1.18 Model of trust based on Blomqvist's concept in dynamic approach.

of relationships between business partners, the present **behavior** (for example, meeting deadlines) becomes more visible and can be more easily assessed. Furthermore, it increases reliability. However, these changes need to be accompanied by **self-reference** to a system that implies capabilities of the company to autonomy and dependence. This system refers to identity, values, targets, and capabilities of the company.

Integration Model of Organizational Trust by Mayer, Davis, and Schoorman (1995)

According to Mayer, Davis, and Schoorman, trust is a very unique feature connected with expectations that others will be worthy of our trust. In their model, they consider propensity to trust (trustor's propensity). It is a stable factor indicating what the probability of trusting the other party is. People are characterized by different aptitudes to trust. This facet is relative to such factors as their personality types, practical experiences, cultural backgrounds, and environments in which they were raised and live. (Hofstede 1980). An example of an extreme case in which this tendency to trust is very high is called "blind trust" ("blind faith"). Certain persons trust in situations where they do not have any reason to. On the other hand, others are unwilling to trust regardless of whether there are very strong premises (Mayer, Davis, and Schoorman 1995). The friendly attitude of a trustor can be described as the propensity of the decision-maker to make or to avoid risky situations (Sitkin and Pablo 1992). However, Mayer, Davis, and Schoorman separate the friendly attitude of a trustor from risk issues. Trust is connected with the perception of risk and then with taking, to some degree, this risk in relationship to the environment: persons, institutions (e.g., bank) and objects (e.g., car). The final result is determined by the level of accepted risk (Figure 1.19).

Like Blomqvist, Mayer, Davis, and Schoorman list **abilities** as the first factor that contributes to the reliability that leads to trust. These abilities are a group of definite skills, talents, or characteristics within a specific field. They always should be considered with reference to some given task that needs to be done. Often, abilities are qualified as competences. Another component of trust is **benevolence,** which is the extent to which a trustee is believed to want good for the trustor. It shows the relationship between trusting parties. In third place is **integrity,** which refers to obeying accepted principles by all parties. All three described components and other examples have been already described in the section "Theoretical Approaches toward Trust."

Different authors cite a different number of elements preceding trust development. The authors of the first two models of trust systems presented here enumerate four elements. Other authors describe trust with ten (Butler 1991), six (Giffin 1967), or only two components (Kee and Knox 1970, Larzelere and Huston 1980, Lieberman 1981). Wekselberg (1996) postulates that the model by Mayer, Davis, and Schoorman should be enriched with additional factors expressing the social character. Opposed, it a poor image of the phenomenon of trust.

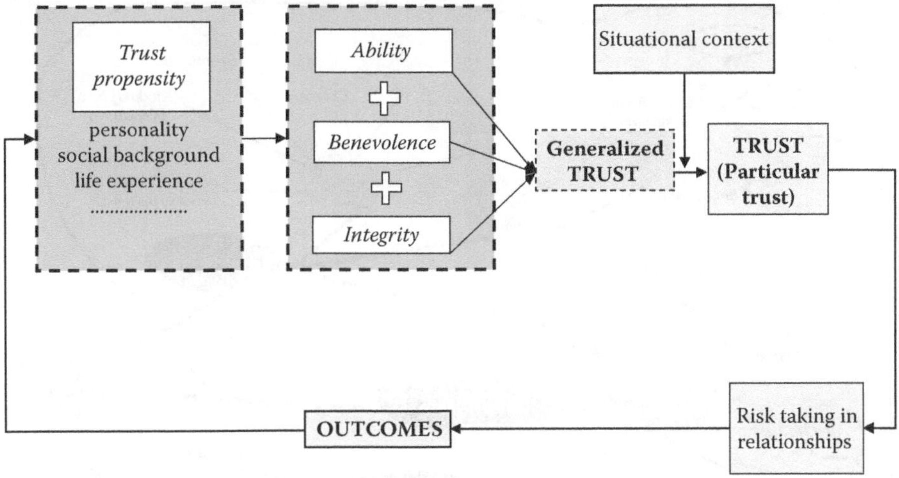

FIGURE 1.19 Model of trust. *Source:* Partly adapted from Mayer, R. C., Davis, J. H., and Schoorman, F. D. 1995. An integrative model of organizational trust. *Academy of Management. The Academy of Management Review,* July 1995; 20, 3; ABI/INFORM Global, p. 715.

Social-Cognitive Trust Model by Cristiano Castelfranchi (2004)

The Social-Cognitive Trust Model by Cristiano Castelfranchi is grounded in the "psychical" image of trust in the context of such cognitive categories as targets and endeavors (Castelfranchi and Falcone 1998, 2000, 2002; Falcone and Castelfranchi 2001). In this model, trust is recognized in three ways:

- Relationship in mental categories (expectation, assessment) to the object of trust (in other words, disposition)
- Decision whether or not to trust can put trustor in danger
- Behavior—for instance, "intuitive" trust based on previous relations among the parties involved in a trust enactment.

Every trust concept connotes that there are various cognitive elements in a trustor's head. A model built on attitude, called belief-desire-intention (BDI), applies the mental condition in modeling (Castelfranchi 2004, p. 4).

The model does not show the full "psychological" dimension of trust, only the most univocal and certain forms (Figure 1.20). The model contains factors connected not only with faith but also with aims. We consider the decision of X whether to trust and to rely on Y. The target is in the model a basic notion, explaining the motivation of all involved parties. Aims are **possibilities** that engender trust to somebody or something. A target is also that this somebody or something **does not injure** us. We intend also to **foresee** what intentions and competences the other party has. In the model, there are also pointed out **competences** as the aims; these competences are tied with self-confidence, abilities, and know-how. Albeit this proposed enlargement (the development) of the model can be changed. However, there stay four first invariable elements and further added elements can be changed depending on needs.

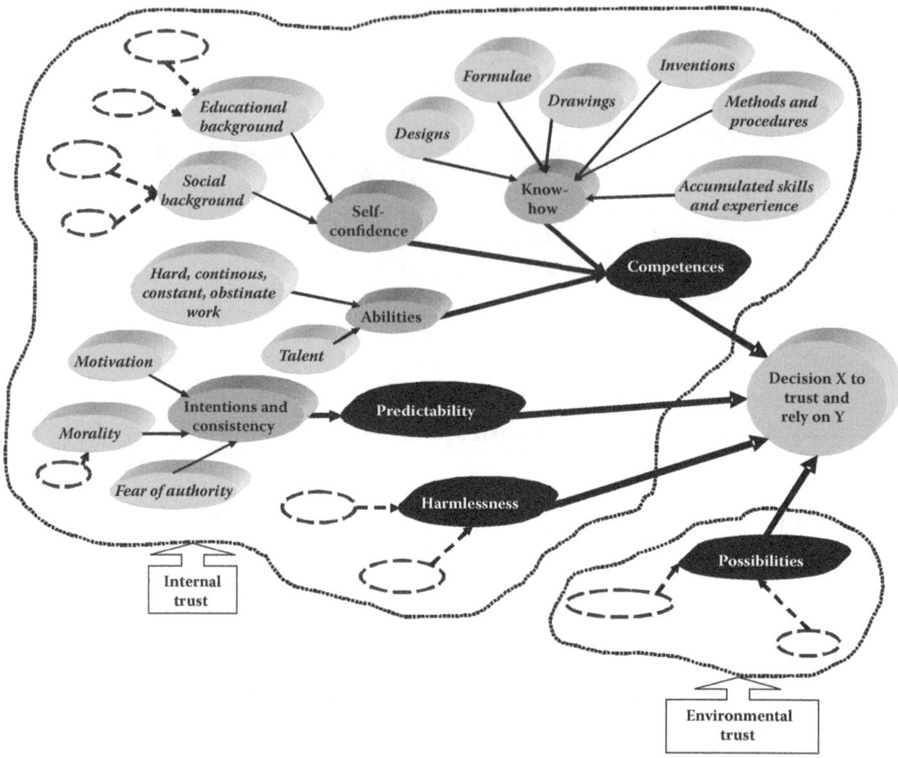

FIGURE 1.20 Social-Cognitive Trust Model by Cristiano Castelfranchi (2004). *Source:* Adapted from Castelfranchi, C. 2004. Trust mediation in knowledge management and sharing. In *Trust management: Second international conference,* C. Jensen, S. Poslad, and T. Dimitrakos (eds.), 304–318. iTrust 2004, Oxford, UK, March 29–April 1, 2004, Proceedings (Series: Lecture Notes in Computer Science), Vol. 2995.

The Virtual Factory (Virtuelle Fabrik) Trust Model According to Lavrac, Ljubic, Jermol, and Bolhalter (2005)

Trust modeling in organizations, also in VOs, results in the necessity of characterizing trust. Indispensable information can be easily obtained by having all partners fill in simple questionnaires. The questions refer to results of the concrete partner and partners of this partner within the framework of previous cooperation. Ergo, every partner fills in the questionnaire, along with all the groups within the organization with which they have cooperated in the past.

Proposed questionnaires should embrace the following three main areas:

- Skills of certain partner (the list of skills)
- Skills of partners of certain partner (the list of skills)
- Estimation of trust of certain partner grounded on the estimated reputation of the partner (image, market share, etc.), the number of finished successful cooperations from the past, estimated profits from the cooperations,

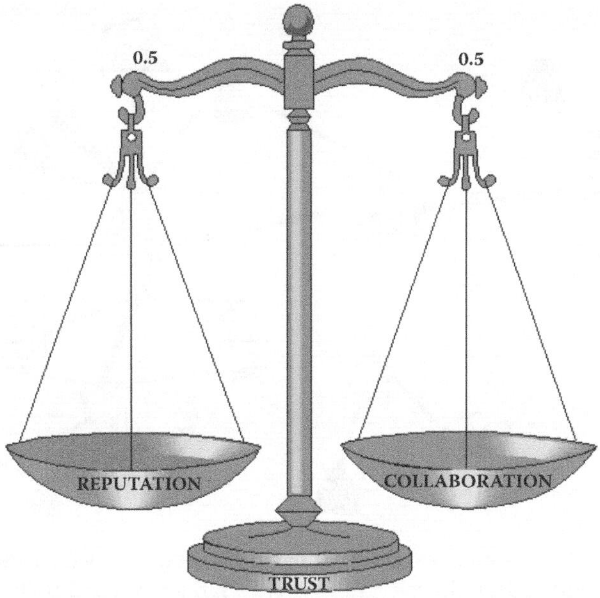

FIGURE 1.21 Trust assessment model. *Source:* Adapted from Lavrac, N., Ljubic, P., Jermol, M., and Bolhalter, S. 2005. A decision support approach to trust modeling in networked organizations. In *Collaborative networks and their breeding environments*, L. M. Camarinha-Matos (ed.). New York: Springer, p. 170.

estimated assignments performed properly and on time, estimated quality of results and products of the partner, estimated expediency of activity costs, and costs of products.

Trust modeling in compliance with the idea of Lavrac et al. demands the estimation of trust. It is allowed to modify the manner of rating trust. The sample model of trust evaluation is presented in Figure 1.21 (Lavrac, Ljubic, Jermol, and Bolhalter 2005).

In compliance with the above model of trust evaluation, we make modeling of the reputation across the following partner's properties: activity, partnership relations, propensity to risk, punctuality, reliability, and economic situation. Each property is assigned a value from 1 to 6 (with 1 representing "very bad" and 6 representing "very good"). "General reputation" is determined by averaging these basic values. Cooperation among partners is rated on a scale of 0 to 3 (with 0 representing "non-cooperation" and 3 representing "strong cooperation").

As a result of explored questionnaire we receive the network which is stellar. In the center of such a network can be found the certain main partner (or partners), and connections from the center outside represent connections between partners. The more connections a main partner has with other partners, the more complicated the network that is created.

Figure 1.22 shows more clearly the best joint organizations, so-called subgroups ("cliques") of partners with a very high degree of trust. The connections look like a web of links between different partners. These links can be very short, simple or

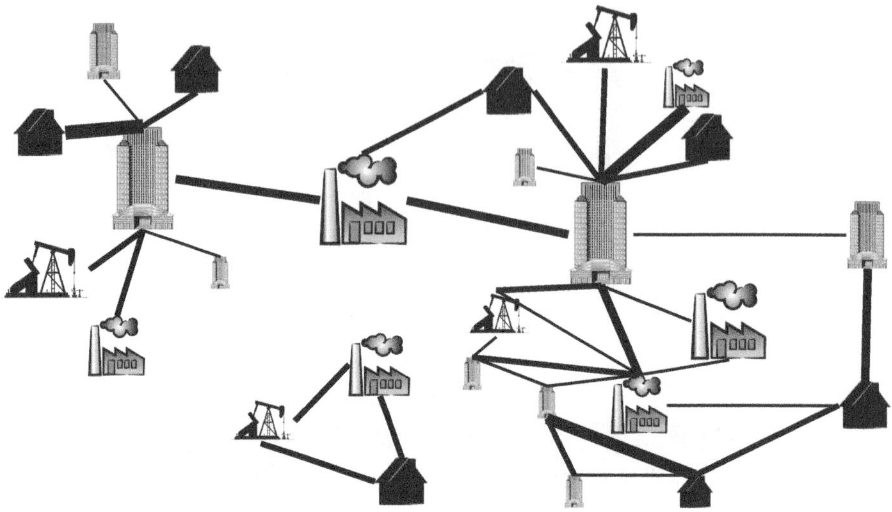

FIGURE 1.22 Visualization of trust modeling in virtual factory with node groups of partners. Source: modified from Lavrac, N., Ljubic, P., Jermol, M., and Bolhalter, S. 2005. A decision support approach to trust modeling in networked organizations. In *Collaborative networks and their breeding environments*, L. M. Camarinha-Matos (ed.). New York: Springer, p. 172.

complex, and involve many partners at the same time. By using a visualization tool to create such figures as the one above, we can evaluate trust level in the Virtuelle Fabrik. The thicker the junction line between partners, the higher the level of trust.

The method of the trust modeling introduced in the Virtuelle Fabrik allows us to conclude the following:

- Fully developed tools helping in the decision-making process are very useful at the evaluation of the trust level.
- The figure representing trust modeling shows the exact state (the situation) for the Virtuelle Fabrik and potential future virtual organizations.
- The analysis of Figure 1.20 allows us to point out authorities in the sphere of trust that can be examples for other partners.
- Strong relationships among partners can be confirmed in practice.
- One ought to collect some exact data, because trust is a very sensitive issue in every organization. Problems can result also from geographical distance in partnership relations or from other factors that make it difficult to gain data needed to model trust.
- In spite of the existing danger of the injury of off-line and non-joint partners, the proposed approach to trust modeling can also have a stimulating effect on off-line partners. It can help to explain such situations and to actuate the will to be with active members of a network.

The proposed approach to trust modeling in VOs can be used for many different aims, such as the general estimation of trust, visualization of the entire trust network, and discovery of very well-joined subnets with a high value of trust that represent so-called "cliques" of partners about the very strong mutual trust. Moreover, it is possible to create rankings of the best partners in VOs, which show authorities in the sphere of trust.

Theoretical Trust Model by Dutton (2000)

At this point, we have presented general models and one particular case of Virtual Factory. However, there exist many models of particular trust. An example is the model described below (Figure 1.23), which can be applied to research on trust in e-commerce and VOs. The process of building trust in electronic business in relations C2B and B2B is described in detail in chapter 2, sections 2.6. and 2.7.

The Theoretical Trust Model built by Dutton consists of four main elements. Elements of the model are derived from Van Slyke and Belanger's (1999) model of e-commerce. The first component is **predictability.** It is built by consequence in activities (Tiller, Holmes, and Zanna 1985) and refers to the ability of the trustor to qualify what the other party will do in a given situation. The second component is **technical competences,** which are especially important in e-commerce. This component deals with devices used in communication to electronic data interchange (EDI) systems, which can use electronic mail or web pages WWW, official records HTTP, or official records of data protection SSL for transmission of the information. The use of EDI within the framework of an organization demands special trust not only in traditional methods of the work but also in these modern (like workflow). Trust is referred not only to people but also to technical devices. It is very important to coordinate all practical systems of data protection (Hart and Saunders 1997). The increase in technical competences entails an increase in trust. On the other hand, the lack of or not large enough technical competences determines in a modern manner a huge barrier in trust building. Certain minimum technical requirements ought to be realized so that trust will develop. **Reliability** is the third component in the model by Dutton. It means that there is concordance in how a person speaks and what he does (Hart and Saunders 1997). Reliability is also described as trustworthiness and agreement between how somebody behaves and how one should behave (Johnson-Georger and Swap 1982). Lack of agreement diminishes our willingness to trust. Last, however not less important from all of the above, is **fiduciary responsibility** reflecting the trustor's situation (Barber 1983; Dutton 2000). This is expressed as a duty of the organization to do the activity in the name of the consumer and for his good, even when he does not have the sufficient knowledge or experience to evaluate the activities of this organization. For example, a lawyer should act in the name of his client no matter whether this client understands the possibility, which is created by existing law regulations. A high degree of fiduciary responsibility in a trusting party causes a rise in the level of trust.

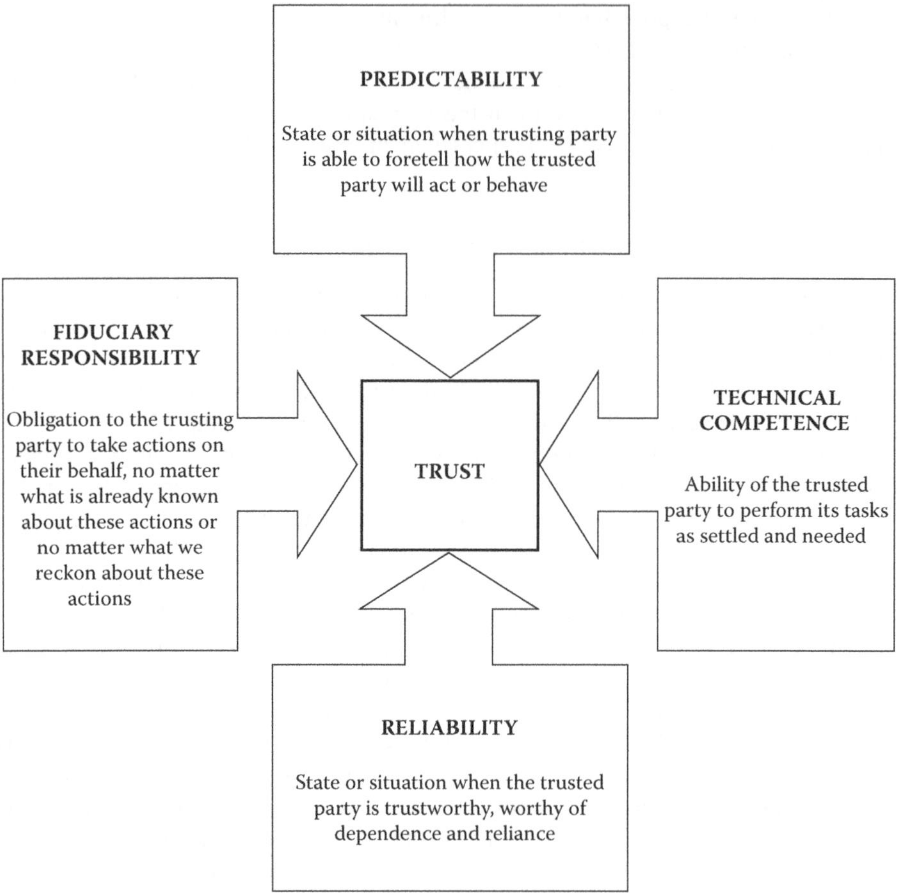

FIGURE 1.23 Theoretical trust model. *Source:* Adapted from P. Dutton, Ph.D. 2000. *Trust: Issues in the design and development of electronic commerce systems.* Griffith University School of Computing and Information Technology, October, and Van Slyke, C., and Belanger, F. 1999. *Trust in consumer-oriented electronic commerce.* Working Paper, 1–17.

REFERENCES

Axentowicz, G., Sankowska, A., and Wańtuchowicz, M. 2006. Modelowanie zaufania w organizacjach wirtualnych. In *Komputerowo zintegrowane zarządzanie,* R. Knosala (ed.). Opole: Oficyna Wydawnicza Polskiego Towarzystwa Zarządzania Produkcją.

Barber, B. 1983. *The logic and limits of trust.* New Brunswick, NJ: Rutgers University Press.

Blomqvist, K. 2002. *Partnering in the dynamic environment: The role of trust in asymmetric technology partnership formation.* University of Lappeenranta, Finland: Acta Universitatis Lappeenrantaensis, 122.

Butler, J. K. 1991. Towards understanding and measuring conditions of trust: Evolution of a conditions of trust inventory. *Journal of Management,* 17, 643–663.

Castelfranchi, C. 2004. Trust mediation in knowledge management and sharing. In *Trust management: Second international conference,* C. Jensen, S. Poslad, and T. Dimitrakos (eds.), 304–318. iTrust 2004, Oxford, UK, March 29–April 1, 2004, Proceedings (Series: Lecture Notes in Computer Science), Vol. 2995.

Castelfranchi, C., and R. Falcone. 1998. *Principles of trust for MAS: Cognitive anatomy, social importance, and quantification.* Proceedings of the International Conference on Multi-Agent Systems (ICMAS'98), AAAI Press, 72–79.

Castelfranchi, C., and Falcone, R. 2000. Trust and control: A dialectic link. *Applied Artificial Intelligence Journal,* Special issue on "Trust in Agents" Part 1, C. Castelfranchi, R. Falcone, B. Firozabadi, and Y. Tan (eds.), 14(8), 799–823.

Dutton, P., Ph.D. 2000. *Trust: Issues in the design and development of electronic commerce systems.* Griffith University School of Computing and Information Technology, October.

Falcone, R., and Castelfranchi, C. 2002. Social trust theory. In *The open agent society,* J. Pitt (ed.). Chichester, UK: John Wiley and Sons, Ltd.

Falcone, R., and Castelfranchi, C. 2002. The socio-cognitive dynamics of trust: Does trust create trust? In *Trust in cybersocieties: Integrating the human and artificial perspectives,* R. Falcone, M. Singh, and Y. H. Tan (eds.). New York: Springer Verlag.

Findeisen, W. (ed.). 1985. *Analiza systemowa – podstawy i metodologia.* Warszawa: PWN.

Giffin, K. 1967. The contribution of studies of source credibility to a theory of interpersonal trust in the communication department. *Psychological Bulletin,* 68, 104–120.

Grudzewski, W. M., Sankowska, A., and Wańtuchowicz, W. 2005. Determinants of business partners acceptance in collaborative working environments, *Economics and Organization of Enterprise,* 6(665).

Grudzewski, W. M., Sankowska, A., and Wańtuchowicz, W. 2006. Trust management in virtual organisations of small- and medium-sized enterprises. *Economics and Organization of Enterprise,* 1(672), 3–10.

Hart, P., and Saunders, C. 1997. Power and trust: Critical factors in the adoption and use of electronic data interchange. *Organisational Science,* 8(1), 23–42.

Hofstede, G. 1980. Motivation, leadership, and organization: Do American theories apply abroad?, *Organizational Dynamics,* 9(1), 42–63.

Johnson-Georger, C., and Swap, W. 1982. Measurement of the specific interpersonal trust: Construction and validation of a scale of assess trust in a specific other. *Journal of Personality and Social Psychology,* 43(6), 1307–1317.

Kaufmann, F., Kokalj, L., and May-Strobel, E. 1990. *EG-Binnenmarkt: Die grenzüberschreitende Kooperation mittelständischer Unternehmen- Empirische Analyse von Möglichkeiten,* Voraussetzungen und Erfahrungen. Stuttgart: C. E. Poeschel.

Kee, H. W., and Knox, R. E. 1970. Conceptual and methodological considerations in the study of trust. *Journal of Conflict Resolution,* 14, 357–366.

Larzelere, R., and Huston, T. 1980. The dyadic trust scale: Toward understanding interpersonal trust in close relationships. *Journal of Marriage and the Family,* 42, 595–604.

Lavrac, N., Ljubic, P., Jermol, M., and Bolhalter, S. 2005. A decision support approach to trust modeling in networked organizations. In *Collaborative networks and their breeding environments,* L. M. Camarinha-Matos (ed.). New York: Springer.

Lieberman, J. K. 1981. *The litigious society.* New York: Basic Book Publishing.

Mayer, R. C., Davis, J. H., and Schoorman, F. D. 1995. An integrative model of organizational trust. *The Academy of Management Review,* July 1995; 20, 3; ABI/INFORM Global, 709.

McKnight, D. H., Cummings, L. L., and Chervany, N. L. 1998. Initial trust formation in new organizational relationships. *Academy of Management Review,* 23, 437–490.

Rempel, J. K., Holmesand, J. G., and Zanna, M. P. 1985. Trust in close relationships. *Journal of Personality and Social Psychology,* 49(1), 95–112.

Sitkin, S. B., and Pablo, A. L. 1992. Reconceptualizing the determinants of risk behavior. *Academy of Management Review,* 17, 9–38.
Van Slyke, C., and Belanger, F. 1999. *Trust in consumer-oriented electronic commerce.* Working Paper, 1–17.
Wekselberg, V. 1996. Reduced "social" in a new model of organizational trust, *Academy of Management Review*; 21(2); ABI/INFORM Global, 333.

1.7 METHODOLOGY OF TRUST MEASUREMENT

A pivotal facet of trust management is its measurement. This section explores approximately the uppermost chosen ideas in this field. When trust is an object of the measurement, we cannot take a homogeneous approach. This results from the multidisciplinary character of trust, which can be deliberated, for instance, from sociological, psychological, marketing, or management sciences points of view. Some authors opine that trust is a one-dimensional variable (Roberts and O'Reilly 1974, Robinson and Rousseau 1994, Tsui et al. 1997), whereas others consider it in several dimensions. Findings show that it is better to apply a multidimensional approach. So the next consideration is how many dimensions we ought to take into account. Cook and Wall (1980) and McAllister (1995) draw attention to two dimensions, Clark and Payne (1997) five, and Butler (1991) eleven. The list of trust dimensions can be never-ending. So the problem is to choose the most meaningful dimensions of trust to measure it. Below is a short introduction to chosen methods and tools in trust research and the approach to trust measurement.

Essential in trust research are accessible methods and tools. Among them we ought to mention:

1. **Psychological games**—stagings of standard, interpersonal situations in organizations that reflect interest conflicts. The usage of interpersonal games in trust research dates back to the 1950s. The most well-known game is "the dilemma of the prisoner." In this game, always most remunerative are confrontation strategies. However, the optimum result—the maximization of economic effects—one can accomplish using the cooperation strategy. In this game exists an assumption that interpersonal trust is the basis for all types of cooperation between game participants (Deustch 1960).
2. **Questionnaire methods**—measuring expectations of the individual in social situations containing univocally expressed statements. Respondents take a position to particular situations, opinions, and the like.
3. **Behavioral measures**—distinguished, observable behaviors of investigated persons on the grounds that the explorer infers about the degree of trust discovered in the experimental (with the produced or spontaneous character) situation. Two main behavioral scales are applied to estimate behaviors of workers: Behaviorally Anchored Rating Scale (BARS) and Behavioral Observation Scale (BOS).
4. **Measurement of interorganizational trust**—relies mainly on estimation of the reliability of interaction's partner. An example of such a tool is "Mindshift" built by Segil (Bratnicki and Strużyna 2001), which consists of

three main stages. In the first stage, one defines the phase of the life cycle in which participants of the cooperation are currently situated. Then every manager responsible for functioning of the given alliance is assigned to one of six types of individuality. The last stage is to determine the type of cooperation project. So formed in that way, the triangle becomes an evaluation base for organizational trust concerning the cooperation.

According to Paine (2003, p. 7), the trust measurement process consists of the following main six phases:

1. **Determining the tested community**—This step is to define the next phases (e.g., choice and form of research tools).
2. **Settling the measurable targets of research**—There is a need to define very precisely what trust covers, which aspects of this phenomenon will be elucidated, and so forth.
3. **Finding the datum point for one's findings**—The character of research should assure the comparative analysis (i.e., we ought to check how trust is being measured in our branch, field, etc. and accept methods that make possible the comparison of our results). Results that do not have a datum point lose their own investigative value.
4. **Choosing the research tool and conducting preliminary (initial) research**—There is no one universal research tool; the information about choosing it depends on previous stages. In the following section, many methods applied in trust research (e.g., questionnaires, focus groups, ethnographic research, models building) are introduced.
5. **Analyzing preliminary results and renewing research.**
6. **Determining trust level for the tested community.**

The above-proposed order of activities during measurement of trust phenomenon is often used in practice in order to decrease failure during the research. It provides a structure for research to be conducted in a well-ordered and harmonious manner.

Cummings and Bromiley (1996)

Cummings and Bromiley (1996) created their own scale called the *Organizational Trust Inventory* (OTI), belonging to and used by certain organizations. This very unique scale contains 62 elements. But, the *Organizational Trust Inventory–Short Form* (OTI–SF) is more common in practice. The OTI-SF is a 12-item Likert-type scale geared toward assessing three dimensions of trust, including reliability, honesty, and good faith in the other party that it will fulfill its obligations.

According to Cummings and Bromiley (Watson 2005), the OTI-SF is as follows:

1. We think the people in ______ tell the truth in negotiations.
2. We think that ______ meets its negotiated obligations to our department.
3. In our opinion, ______ is reliable.

4. We think that the people in ______ succeed by stepping on other people [reverse coded].
5. We feel that ______ tries to get the upper hand [reverse coded].
6. We think that ______ takes advantage of our problems [reverse coded].
7. We feel that ______ negotiates with us honestly.
8. We feel that ______ will keep its word.
9. We think ______ does not mislead us.
10. We feel that ______ tries to get out of its commitments [reverse coded].
11. We feel that ______ negotiates joint expectations fairly.
12. We feel that ______ takes advantage of people who are vulnerable [reverse coded].

Larzelere and Huston (1980)

Larzelere and Huston (1980) built the Dyadic Trust Scale (DTS) for trust measurement basing on the previous studies. This scale was created to understand interpersonal trust in very close relationships. They defined trust as dyadic because they believed "trust exists to the extent that a person believes another person (or persons) to be benevolent and honest" (p. 596). Larzelere and Huston recognize benevolence as a significant component of dyadic trust. Benevolence is described as whether a person is motivated by his or her own desires, or if he or she is motivated by gain in the relationship jointly with the cooperation participant. Although the authors began with over 40 items, the final DTS is built from the following eight-itemed, seven-pointed Likert-type measurement (Larzelere and Huston 1980, p. 599), which can be grouped in two major areas:

1. **Concerning partner**—we assess whether:
    - our partner is primarily interested in his or her own welfare
    - our partner is perfectly honest and truthful with us
    - our partner is truly sincere in his or her promises
    - our partner treats us fairly and justly
    - there are times when our partner cannot be trusted
2. **Exploring ourselves:**
    - We can trust our partner completely
    - Our partner does not show us enough consideration
    - Our partner can be counted on to help us

The Dyadic Trust Scale is used at Child Trends, "which is a nonprofit, nonpartisan research organization dedicated to improving the lives of children by conducting research and providing science-based information to improve the decisions, programs, and policies that affect children and their families" (http://www.childtrends.org). They use the scale created by Larzelere and Huston (1980) under the heading "Healthy Marriages Compendium" (Child Trends n.d.).

McAllister (1995)

McAllister (1995) believed that there are two types of trust foundations: one grounded in cognitive judgments of another's competence or reliability (referred to as *cognition-based trust*) and another based on affective bonds among entities (referred to as *affect-based trust*). According to McAllister, development of cognition-based trust is correlated with affect-based trust. What is more, affect-based trust in turn has a meaningful impact on need-based monitoring and citizenship behavior. These factors were both significantly related to the performance levels of the managers taking part in this study.

McAllister's measure to assess affect- and cognition-based trust levels consists of 11 Likert-type items, 6 assessing levels of cognition-based trust and 5 assessing affect-based trust (described below).

Affective foundations for trust consist of the emotional bonds between individuals, what allows them to build a sharing relationship. People preferably want to feel that: "We can both freely share our ideas, feelings, and hopes. I can talk freely to this individual about difficulties I am having at work and know that (s)he will want to listen" (McAllister 1995, p. 37). It is worth mentioning that very often people make emotional investments in trust relationships and express genuine care and concern for the well-being of their partners that is so strong that it would be a real loss "if one of us was transferred and we could no longer work together" (McAllister 1995, p. 37). Furthermore, people believe in the fundamental virtue of such relationships. There exists a strong belief that these sentiments are repaid somehow, for instance, we expect that when we share our problems with somebody this person will "respond constructively and caringly" (McAllister 1995, p. 37).

The second aspect is the **cognitive** one. It originates in our decisions and choices—specifically, who we will trust in which respects and under what circumstances. We make decisions about certain individuals and coworkers considering existing reasons, looking at past experiences, and looking for good evidence of trustworthiness. According to McAllister, we expect people to be dedicated, competent professionals who are perfectly prepared to do their job. Moreover, we want other work associates to consider the same individual trustworthy.

Statements constructed by McAllister required respondents (managers) to point out their agreement and coherence with 11 different statements about a particular peer at work.

Delgado-Ballester, Munuera-Aleman, and Yagoe-Guillin (2003)

Measurement of trust is very significant in marketing. It is connected with understanding a consumer-brand relationship. This understanding also requires an analysis of the consumer's trust in the brand. This issue is well described by the nowadays' current brand practices and the methods used by brand managers to take advantage of this idea when defining their brands. According to Delgado-Ballester, Munuera-Aleman, and Yagoe-Guillin (2003 pp. 335–353), brand trust can be defined as "the confident expectations of the brand's reliability and intentions in situations entailing risk to the consumer." On the basis of this definition and a review of marketing and brand literature and interviews, the authors developed the Brand Trust Scale (BTS).

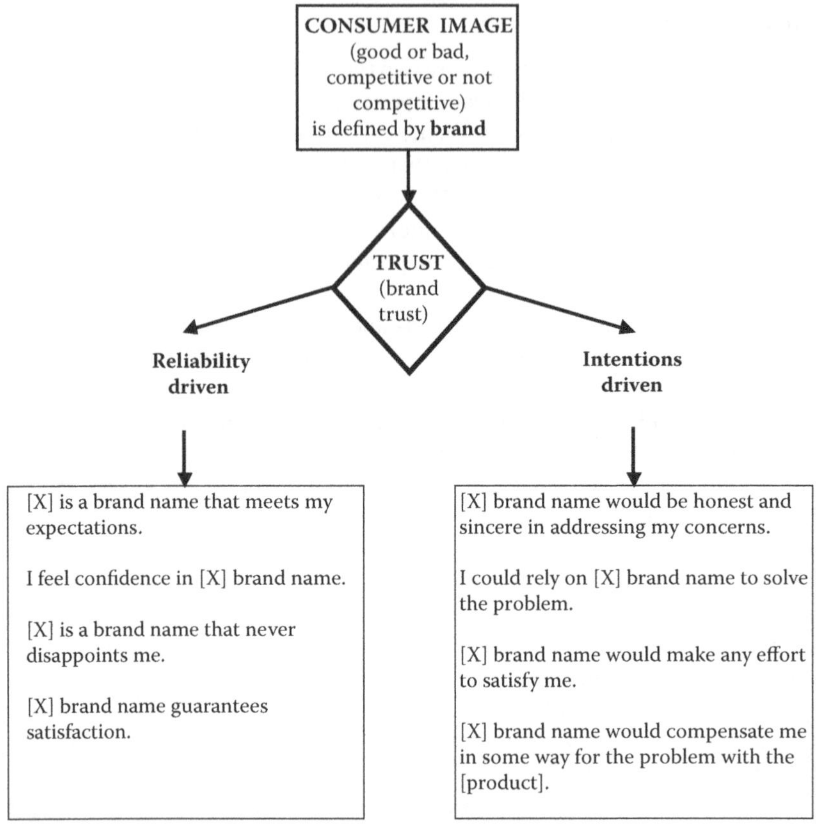

FIGURE 1.24 Conceptual model based on the Brand Trust Scale. *Source:* Adapted from Delgado-Ballester, E., Munuera-Aleman, J. L., and Yagoe-Guillin, M. J. 2003. Development and validation of a Brand Trust Scale. *International Journal of Market Research,* 45(1): 335–53.

They assumed that brand trust evolves from cumulative previous experience, not only from the last act of buying and consumption. They believe that, in this situation, it is necessary to measure overall satisfaction with certain statements (Figure 1.24).

Zaheer, McEvily, and Perrone (1998)

Zaheer, McEvily, and Perrone (1998) studied interorganizational relationships in a way dissimilar to Cummings and Bromiley's (1996) broad organizational study. The notion of interorganizational trust (also known as *external trust*) is defined by them as follows: "the extent of trust placed in the partner organization by the members of a focal organization" (p. 143). This type of trust can enhance and enrich organizational relationships in a variety of contexts, including firm/client, marketing channels, joint ventures, and internal cooperative alliances (Huff and Kelley 2003). Zaheer and colleagues built a scale using a combination of previous scales supplemented with new ideas. An example of an item from this scale is as follows: "Supplier X has always been evenhanded in its negotiations with us, my contact person is trustworthy, and

I would feel a sense of betrayal if my contact person's performance was below my expectations" (p. 157). Their research revealed that interpersonal and interorganizational trust operate in quite different ways. This finding implies that the effects of trust in the interorganizational context are distinct at both the individual and organizational levels of analysis.

Lavrac, Ljubic, Jermol, and Bolhalter (2005)

A very interesting approach to trust measurement is presented by Lavrac, Ljubic, Jermol, and Bolhalter (2005), who are the authors of the trust modeling for the Virtual Factory described in the previous section. Their model consists of two areas: cooperation and reputation. The variable "cooperation" measures the power of relationships between partners on a scale from 0 to 3 (with 0 representing "noncooperation," and 3 representing "very strong cooperation"). The theoretical variable "reputation" includes such elements as activity, punctuality, reliability based on partnership business relation, propensity to the risk, and the economic situation. These elements are measured on a scale from 1 to 6 (with 1 representing "very bad," and 6 representing "very good"). An interesting aspect of this tool of trust measurement between organizations is an attempt to define the elements constituting reputation.

Marcisz (2002)

The essence of trust and its measurement from the information technology (IT) viewpoint is presented in the book entitled *Self-Creating Language* (2002) written by Marcisz. The theory explained in this book is called Secral. The author accepts that trust is always connected with the loss of certain information generally widely available in the environment (Marcisz, 2002; Figure 1.25). From an enormous amount of information reaching us, we select only part.

Deliberating Secral theory can be done on the example of a society. Components of this society are people. We can assume that in the very beginning of observation people contact each other quite steady, everyone with everyone from time to time (Marcisz 2003). Lines are pointing out the information flows.

The society illustrated in Figure 1.25 consists of single elements, human beings; there are no other elements. However, we can easily notice that some elements contact other chosen elements more often than some others. This represents the next stage of society development (Figure 1.25, Phase 2). We can also easily observe thickened lines, which represent often-repeated relations between elements of the society. The longer observation permits separation of nascent social groups. Within the framework of the concrete group, mutual relations are intensified very much.

Previously, we presented the manner of informational grouping of objects. This method refers not only to society but also to other objects, both alive and inanimate. Within the framework of separated objects (i.e., the social group), the exchange of information among elements is varying. So the structure of the social group is not homogenous. Explanation of the trust-building process is necessary to present the original idea of trust measurement in compliance with an IT approach.

Trust built in the above-presented manner can be measured in two ways. The first method consists of indirect observation of symptoms. Members of the given

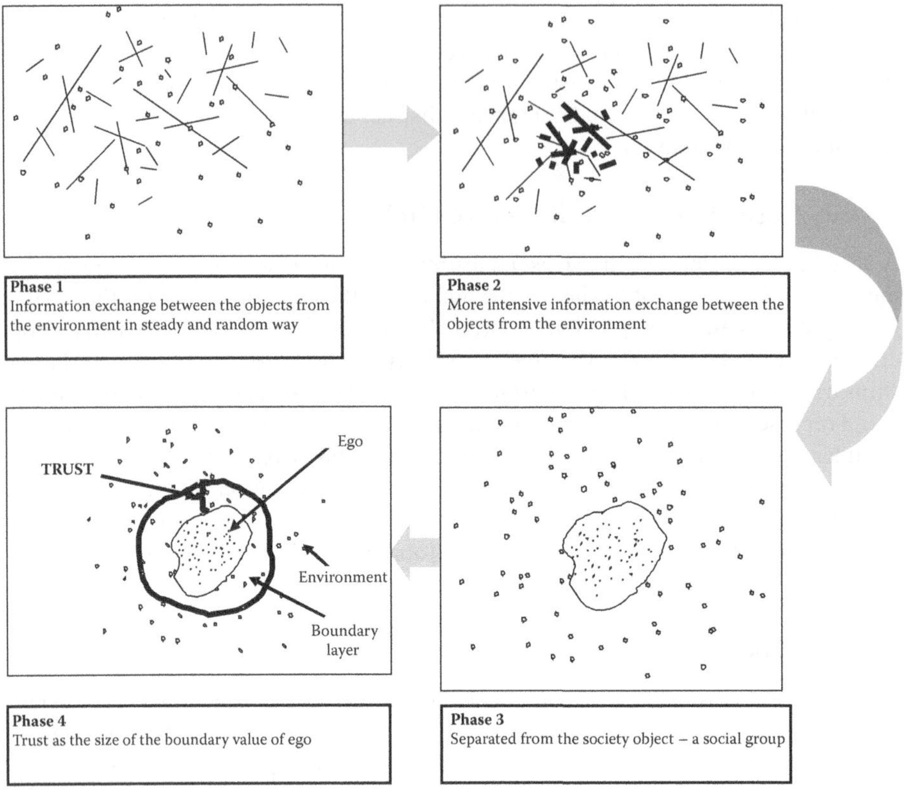

FIGURE 1.25 Trust formation and measurement in society: Conceptual model. *Source:* Adapted from Marcisz, A. 2003. Istota zaufania i jego pomiar. *Prakseologia*, 143/2003, p. 154–157.

social group are handed questionnaires and tests. An imperfection of this method is its sensibility to many factors of the psychological and social nature not connected with the research object (Alba and Hutchinson 2000; Luce, Jia, and Fischer 2002).

The second method is direct measurement based on the inquiry description of trust. The trust is equal to self-identification with other objects and to the fluent information exchange between objects (Marcisz 2003, p. 157). Ego includes objects being characterized with very strong information exchange. In the boundary layer are found objects with the smaller intensity of the information exchange. The boundary layer of ego separates the object from the rest of the environment. The width of the boundary layer of ego testifies to the trust level of ego to the rest of the environment and to the intensity of the information flow. The narrow layer means that there is a strong and univocal distinction between the interior of the object and the environment. The wide layer is the sign that the object easily accepts elements from the environment.

The second method assumes that the measurement refers to the cumulative frequency of the information transmission. There exist two possibilities: first, symptomatical (indirect), and the second direct. In the case of measurement of symptomatical information transfer, there is not one universal, comprehensive method. Otherwise, we would be measuring trust among separate human beings rather than between all social groups.

We can consider the example of the single man. His border zone consists of things closest to him: persons, places, objects, ideas, and so forth. The quantity of objects forming this zone has no meaning, because we investigate the intensity of the information flow. The trust measurement consists in measuring the size of the informational exchange between the man and other mentioned objects. The greater trust, the less routine assignments. The enlargement of trust expressed by the man is visible in the form of an increased ease of contacts made with other people, other objects, removes, and so forth. Trust is proportional to the quantity of objects and intensities of contacts of the man with these objects.

Direct measurement is more difficult to conduct and closely depends on the degree of skills at the information flow measurement. Information is treated as a physically measurable quantity (Marcisz 2003). So one can measure trust in a similar manner. However, we ought to remember that this way we can measure only its physical image.

In spite of its limitations, the informational approach to trust measurement is a very important step. However, further research on the concrete environment and its objects is necessary.

Trust measurement is all-important in trust management. There exist many methods in this area. Albeit there are still not unbound certain matters, such as whether trust should be measured only from the perspective of single man or single organization or from the perspectives of many organizations simultaneously? Moreover, a controversial matter remains: what is the optimum quantity of criteria in order to be able to describe trust relatively straight and simultaneously as full and precise as possible? The answers to these example questions are the next step toward the permanent consensus on trust measurement within the scientific field. There are still needed further activities in this domain, both of the practical and theoretical character.

REFERENCES

Alba, J. W., and Hutchinson, J. W. 2000. Knowledge calibration: What consumers know and what they think they know. *Journal of Consumer Research*, 27,(2), 123–156.

Bratnicki, M., and Strużyna, J. (eds.). 2001. *Przedsiębiorczość i kapitał intelektualny*, Katowice: Wydawnictwo AE w Katowicach.

Butler, J. K. 1991. Toward understanding and measuring conditions of trust: Evolution of a conditions of trust inventory. *Journal of Management*, 17, 643–663.

Child Trends. n.d. *Trust measures*. http://www.childtrends.org/Files/21_TrustMeasures_web.pdf (accessed June 30, 2006).

Clark, M. C., and Payne, R. L. 1997. The nature and structure of workers' trust in management. *Journal of Organisational Behaviour*, 18, 205–224.

Cook, J., and Wall, T. 1980. New work attitude measures of trust, organizational commitment and personal need non-fulfillment. *Journal of Occupational Psychology*, 53, 39–52.

Cummings, L. L., and Bromiley, P. 1996. The Organizational Trust Inventory (OTI): Development and validation. In *Trust in organizations: Frontiers of theory and research*, R. M. Kramer and T. R. Tyler (eds.). Thousand Oaks, CA: Sage Publications.

Delgado-Ballester, E., Munuera-Aleman, J. L., and Yagoe-Guillin, M. J. 2003. Development and validation of a Brand Trust Scale. *International Journal of Market Research,* 45(1), 335–53.

Deustch, M. 1960. The effect of motivational orientation upon trust and suspicion. *Human Relations,* no. 13.

Huff, L., and Kelley, L. 2003. Levels of organizational trust in individualist versus collectivist societies: A seven-nation study. *Organization Science,* 14(1), 81–90.

Larzelere, R., and Huston, T. 1980. The dyadic trust scale: Toward understanding interpersonal trust in close relationships. *Journal of Marriage and the Family,* 42, 595–604.

Lavrac, N., Ljubic, P., Jermol, M., and Bolhalter, S. 2005. A decision support approach to trust modeling in networked organizations. In *Collaborative networks and their breeding environments,* L. M. Camarinha-Matos (ed.). New York: Springer.

Luce, M. F., Jia, J., and Fischer, G. W. 2002. *Do I know whether I like it?: Extending work on within-alternative conflict to measures of confidence in consumer judgements,* http://mktg-sun.wharton.upenn.edu/ideas/pdf/01-010.pdf.

Marcisz, A. 2002. *Self-creating language.* Victoria, BC: Trafford.

Marcisz, A. 2003. Istota zaufania i jego pomiar. *Prakseologia,* 143, 153–160.

McAllister, D. J. 1995. Affect- and cognition-based trust as foundations for interpersonal cooperation in organizations. *Academy of Management Journal,* 38, 24–59.

Paine, K. D. 2003. *Guidelines for measuring trust in organizations.* The Institute for Public Relations.

Roberts, K. H., and O'Reilly, C. A. 1974. Measuring organizational communication. *Journal of Applied Psychology,* 59, 321–326.

Robinson, S. L., and Rousseau, D. M. 1994. Violating the psychological contract: Not the exception but the norm. *Journal of Organizational Behavior,* 15, 145–159.

Tsui, A. S., Pearce, J. L., Porter, L. W., and Tripoli, A. M. 1997. Alternative approaches to the employee-organization relationship: Does investment in employees pay off? *Academy of Management Journal,* 40, 1089–1121.

Watson, M. L. 2005. *Can there be just one trust? A cross-disciplinary identification of trust definitions and measurement.* Institute for Public Relations. http://www.instituteforpr.com/ (accessed October 30, 2006).

Zaheer, A., McEvily, B., and Perrone, V. 1998. Does trust matter? Exploring the effects of interorganizational and interpersonal trust on performance. *Organization Science: A Journal of the Institute of Management Sciences,* 9(2), 141–159.

2 Main Assumptions and Principles in Building Trust Systems

2.1 PRINCIPLES OF TRUST BUILDING

The rules of trust creation refer to rules and guidelines that have a far-reaching influence on the formation and development of trust. Trust building is a kind of management strategy because it is strongly focused on not only present but, first of all, future cooperation. The basic guidelines concerning trust building are characterized below (Bibb and Kourdi 2004, p. 161–167):

1. **Self-trust is a critical factor in creating trust.**
   Trust in our own possibilities is a pivotal facet. It is improbable that others will trust in us if we do not trust ourselves.
2. **Waiting until other people prove their own trustworthiness is inadequate (not enough).**
   Trust is not only what is obtained when some people trust us, but we also must trust other people. Mistrust is followed by mistrust and vice versa.
3. **Trust creating and building is a very long process yet trust can be destroyed in one moment.**
   Building trust takes a lot of time, demands engagements, and requires many activities; however, a single incident can strain and even destroy it.
4. **You will not build trust if you treat it as a means to the target.**
   We cannot build trust in a cynical way, because the chances of success are small. People who succeed in trust building genuinely care for their relationships, keep promises, and value frankness and honesty.
5. **We cannot build trust without taking into consideration the values.**
   The values of frankness and honesty are the foundations for trusting relationships and cultures with high trust.
6. **Trust is often invisible.**
   We often do not realize the fact that matters are going on well because of the presence of trust. Only abuse of trust or destruction of trust makes this fact perceptible. Then usually the restoration of trust is already not possible or possible in a very limited range.
7. **Trust demands engagements.**
   Trust does not happen simply. It demands engagements and personal responsibility.

8. **Trust relies usually on reciprocal relations.**
   Trust concentrates on a circle of interaction, "of giving and of gaining." People with well-balanced psyches feel a state of the satisfaction from winning somebody's trust. When we win bigger trust, our propensity to trust usually grows, too.
9. **Trust opens possibilities that are not possible in other cases.**
   Without trust, people and organizations would not be able to be fully effective and creative or achieve successes because trust encourages people to look for new solutions, have differing opinions, and openly express our own opinions even if they will be ridiculed.
10. **Trust changes in time.**
    The phenomenon of trust changes over time. Constant change can cause different complications. A firm in which customers have trust that does not take into consideration its own workers can perniciously acknowledge that this is not a weak spot, the potential source of trouble. Nevertheless, trust and cynicism are like water and get quickly to different parts of the system (the organization). Cynicism is especially corrosive because problems with workers are quickly transferred to customers.
11. **We ought to avoid substitutes of trust.**
    There is no ideal substitute for trust. Formal contracts cannot be found as ideal substitutes of trust. In certain cases, they can bridle trust.
12. **We ought to avoid the games of authority.**
    There appears a natural tendency to take greater part in the chase for authority than in the chase for building trusting relationships. It is the consequence of human ego. Authority assures the inspection, safety, and domination of the human ego. Trust is in turn deeper rooted, more elastic than authority. It acts considerably better. Authority entails domination, and this is a natural antithesis of trust.

We can compare relations based on authority and trust. Firstly, the pursuit of authority means concentration on the business, whereas trust building means acknowledgement that people can accomplish more through cooperation (Bibb and Kourdi 2004, p. 164). Moreover, authority is connected with only one-way communication, creation of a fear atmosphere, and domination as well as autocratism (a conviction about always being right). At the same time, trusting relations are built on two-way communication, understanding, support, reaching consensus, and "fair" behavior. Situation "win-lost," the support of the own elasticity using others' resignations is characteristic of authority-based relations, whereas trust demands mutual engagement and elasticity. Trust demands endeavors to have mastery and understand others, whereas authority requires endeavors to the mastery. Building a trusting atmosphere, in contrast to authority, is possible when we are open to others, parties in prolonged relations are brought nearer, and we are not afraid to depend on others. The influence on others in trusting relations originates from expertise, respect, and understanding; constraint becomes not indispensable. At the same time, authority is always associated with influence on others through constraint. In order to reduce conflict in authority-based organizations, there are built detailed regulations, monitoring is a continuous activity, and legal threat is very commonly used. Trust implies that conflicts are solved by cooperation with people who believe in similar values, by enlarging mutual sympathy, and finally by mediation or arbitration.

13. **We ought to build a culture of high trust in the organization.**
The construction of a trust culture is a key for building trusting relationships. We can list several characteristics of trust culture that create conditions in which people can trust and win somebody else's trust. These are:
    - **Common values.** These values are not only adopted at work, but there is also a basic need for them in everyday life. Common values do not have to necessarily directly refer to worksites but can also refer to communities and environments wherein the individual works.
    - **Common mission and targets.** People must pursue the same aims and be committed to their realization.
    - **Open and authentic leadership.** It is not possible to create an atmosphere of trust if the leader does not trust and is not bestowed trust.
    - **Culture of consensus and not constraint.** Constraint entails the necessity of power exertion to realize some assignments. Trust liberates the intrinsic motivation to conduct these assignments, which is more permanent and engages in the greater effort.
    - **Atmosphere of play and pleasure.** The most innovative firms are those in which people experience satisfaction and joy from their activities, do not fear being ridiculed and, as a result, are more open to look for new ideas and experimentation. Joy at work is a sign that the organizational culture is a culture of trust and frankness (as opposed to dogmatism).
    - **Wish of learning and not blaming.** In a culture focused on finding guilt, people are not willing to trust others and are not so open. If people discover that every adversity will be attributed to them, trust will be dissipated, or it will never be built.
    - **Honest and authentic communication.** Communication is the critical element in trust building. Without effective communication, we always will end up with misunderstandings and concealment of true information. An example is "the syndrome of the nursery school," in which workers get an e-mail or a memorandum about downsizing and are not given the true premises of such a decision (usually the cause is reduction of costs). Acting this way, we do not treat workers as adults, although ironically all surmise the true reasons of dismissals, the hidden information. Trust is built on and remains in the process of authentic communication, even if it is negative (Bibb and Kourdi 2004, p. 121–131).
14. **We ought to avoid the continuous choke of conflicts.**
The entire avoidance of conflicts is groundless based on the viewpoint of its negative impact on the effectivity of relationships and, consequently, on their productivity. However, an optimum level of conflict assures information

exchange and also ensures that an organization can cope with appearing problems. The avoidance of confrontation generates hidden cynicism.

15. **We ought to care for concordance, avoid the strategy of fraud (falsification, concealment, and double meanings) in communication processes.** Trust depends on the concordance of words with activities and behaviors. Several possibilities can be directly interpreted as indications of a discrepancy or lack of credibility—for example:
    - Smiling only with the mouth, not with the eyes
    - Frequent flickering of the eyelids and enlarged pupils
    - Uncertainty of speech (clumsy pauses, stuttering)
    - Maladjustment of body language to expressed and spoken words
    - Words that do not fit the tone of the statement
    - Concrete activities that do not reflect declared rules, convictions, or words.

    Buller and Burgon (cited in Griffin 2003, p. 120–121) use the notion "comma" speaking about unaware nonverbal signals that unveil the internal state of the person sending them. Theoreticians of communication distinguish four characteristic features of announcements that are telltale signs of intended deception:
    - **Uncertainty and lack of precision.** Peter's wish to hide the truth about the Saturday adventure will cause him to answer all questions very generally and evasively. The answer, "I worked/a.m. till late hours" excludes details and causes that with difficulty we can criticize or find something wrong in Peter's behavior. Another typical dodge is the use of passive voice or impersonal forms in the content of the announcement (e.g., "It was difficult to handle everything earlier").
    - **Lack of engagement, taciturnity, and hiding in an "ivory tower."** If Eva appears suddenly and wants to know the reason for her partner's absence at home overnight, the partner thinks mainly how to hide himself somewhere. Such thirst for escape is often reflected in nonverbal activities, such as turning around, making coffee, sitting down further than usual, or sitting up straight rather than leaning forward when giving out answers. We can also expect moments of silence before getting the answer and frequent pauses during speech.
    - **Outdistancing oneself.** Outdistancing is a method of distancing ourselves from our actions. It can appear with the removal of the element of personal choice ("**all** do so") from the announcement, suggesting group responsibility ("**all together** we made it"), and the selection of words to reduce the unpleasantness of conveyed news ("**once in a while,** I am late to work").
    - **Protection of the image and ties.** When deceiving people, we often endeavor to extinguish the reactions of our body that could sign the attempt of deception (for instance, trying to give the impression of great sincerity by frequently nodding assent to the speaker, not interrupting the speaker, and continually smiling).

16. **We ought to reduce uncertainty.**
Reducing uncertainty will result in greater predictability of the behaviors of both parties in an interaction. Along with reducing the ability of the individual to predict which option will now follow, the level of uncertainty is increased. Man has an innate need for some expectations and explanations. Strategies for dealing with initial uncertainty are delivered with axioms by Charles Berger (cited in Griffin 2003, p. 166–168), which are as follows:
Axiom 1: In view of the high level of uncertainty occurring during the beginning of the meeting, due to the increasingly intense verbal communication among strangers it is followed by the fall of the uncertainty level at every from participants of the interaction. A reduction in uncertainty is followed by further intensification of verbal communication.
Axiom 2: Due to more intense, friendly nonverbal expression, the uncertainty level caused by the situation of the first interaction is reduced. The reduction in the uncertainty level enables further intensification of friendly nonverbal expression.
Axiom 3: A high level of uncertainty is the reason actions connected with searching for information are more intense. Together with the fall of the uncertainty level, fewer behaviors connected with information searching occur.
Axiom 4: A high level of uncertainty in the relationship triggers a drawdown of the intimacy characterizing the content of the communication. Low uncertainty level leads to a high level of intimacy.
Axiom 5: A high level of uncertainty contributes to the extent of reciprocity. A low level of uncertainty causes a decrease in reciprocity.
Axiom 6: Similarities between participants diminish uncertainty, and differences enhance uncertainty level.
With the acknowledgement of the reliability of axioms, we can pair them; as a result, we obtain additional insight into the dynamics of relations. From the system of axioms, we can weed out unambiguous statements by means of a pattern of deductive inference:

If A = B
and B = C
then A = C

Berger conducted the above operation for every possible pair of axioms, receiving 28 theorems. For example:

If resemblance decreases the level of uncertainty (axiom 6),
And decreased level of uncertainty causes a rise in sympathy (axiom 7),
Then among resemblance and sympathy exists affirmative dependence (Theorem 21).

17. **We ought to remember that observable behaviors represent "only the tip of the iceberg."**
Certain powers placed below the tip of the iceberg are not directly observable but still affect human behaviors (Figure 2.1). This does not mean that people purposely "abuse trust." There can be many reasons, including unconscious ones, that affect the ability of an individual to build trusting relationships, for example:
    - People might have a strong need to be approved of and popular.
    - They might feel uncomfortable without certainty in a situation when they have low self-assessment.
    - They might want to adapt themselves to behaviors that are perceived as necessary for successful widespread praise.
    - They might have a strong need for authority and domination over others.
    - They might have a strong need to protect their own ego.

    Familiarity with the above-mentioned reasons is important in trust building across organizational culture and leadership.
18. **We ought to treat other people honestly and justly.**
We ought to comply with distributive and procedural justice. People treated justly have greater trust. Distributive justice bears on the manner of awarding a prize. People feel that they are treated justly, if they believe that they obtained an award that corresponds to their contribution and that they received, what was promised to them, and what they need (Figure 2.2).

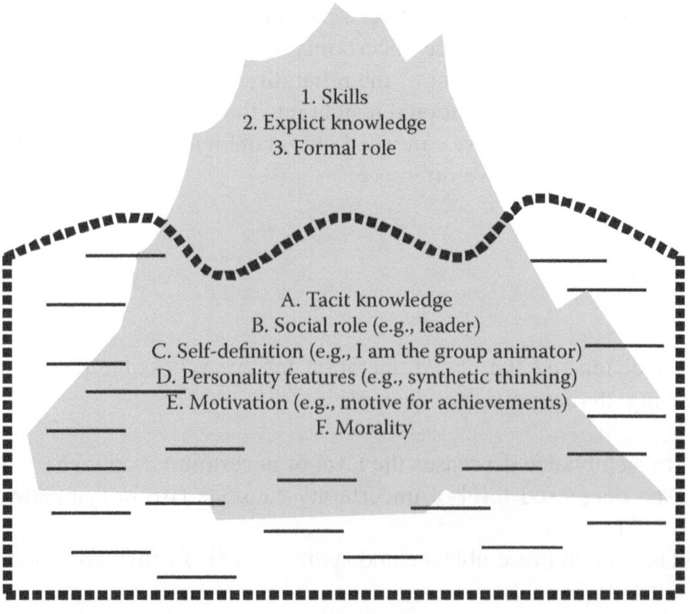

FIGURE 2.1 "Iceberg" of employees' competencies.

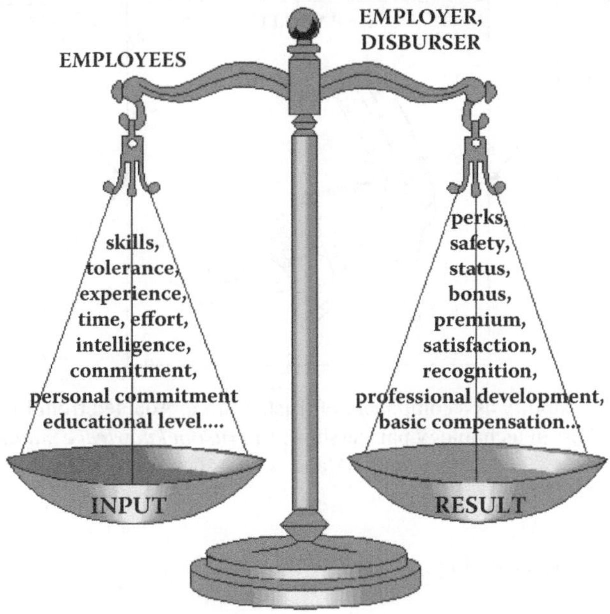

FIGURE 2.2 Distributive justice according to Adams (1965). Source: Adapted from Adams, J. S. 1965. Injustice in social exchange. In *Advances in experimental psychology*, Vol. 2., L. Berkowitz (ed.). New York: Academic Press.

Procedural justice relates to the way of making decisions by management and managing personal procedures. Tyler and Bies (cited in Armstrong 2005, p. 274–276) list five factors that influence the perception of procedural justice:

- taking into account the worker's point of view
- eliminating personal anticipations and stereotypes
- applying identical criteria to all workers
- giving workers quick feedback concerning results of certain decisions
- providing workers with exhaustive explanations of the subject concerning undertaken decisions.

When building trusting relationships, we can focus on strategies explaining how to form each of the elements. Referring to the trust model built by Blomqvist, which was presented in the section "Modeling Trust Systems" in Chapter 1, we can prepare a list of elements enacting trust across the formation of abilities, self-referencing, good will of the enterprise (the institution), and behavior. Figure 2.3 to Figure 2.6 present this idea in more detail.

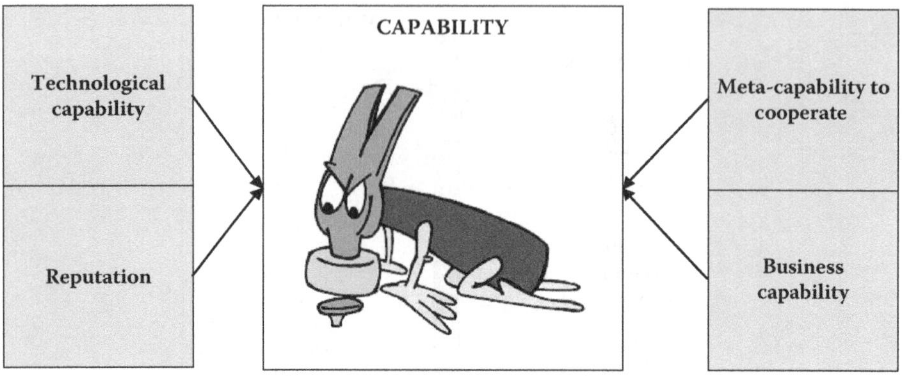

FIGURE 2.3 Capability as a component of trust. Source: Adapted from Blomqvist, K., and Pirjo, S. 2004. Trust in technology partnerships. In *Trust in knowledge management systems in organizations*, M.- L. Huotari (author) and M. Iivonen (ed.). Hershey, PA: Idea Group Publishing, p. 189–192.

Capability is described by:

- **Meta-capability to cooperate**—first-to-market technologies, third-party screening, symbols, realistic judgment, soundness of vision, professionalism, clear strategy and business plan, partner strategy, program and managers, internal information on status and purpose of partnership
- **Business capability**—prizes for outstanding technology
- **Reputation**
- **Technological capability**—references of difficult projects.

Self-reference (clarity of the individual and organizational identity) consists of:

- **Integrity**—consistency of boundary spanners and organization, organizational social control, and sanctions
- **Equity**—sense of fairness (input/output)
- **Double contingency**—negotiation style (win-win), choice of boundary spanners (with experience/understanding from both worlds), project champions, interdependence at project level, self-confidence, and corporate identity
- **Security and stability**—consistency of values and behavior
- **Ability to understand difference and appreciate complementarity**—internal analysis of own competencies and evaluation of needed complementarity, positive attitude to cooperation.

Good will (moral responsibility and positive intentions toward others):

- **Reliability**—norms
- **Care and concern**—consideration for mutual needs, proactive information, interest and advice, social support
- **Organizational culture**

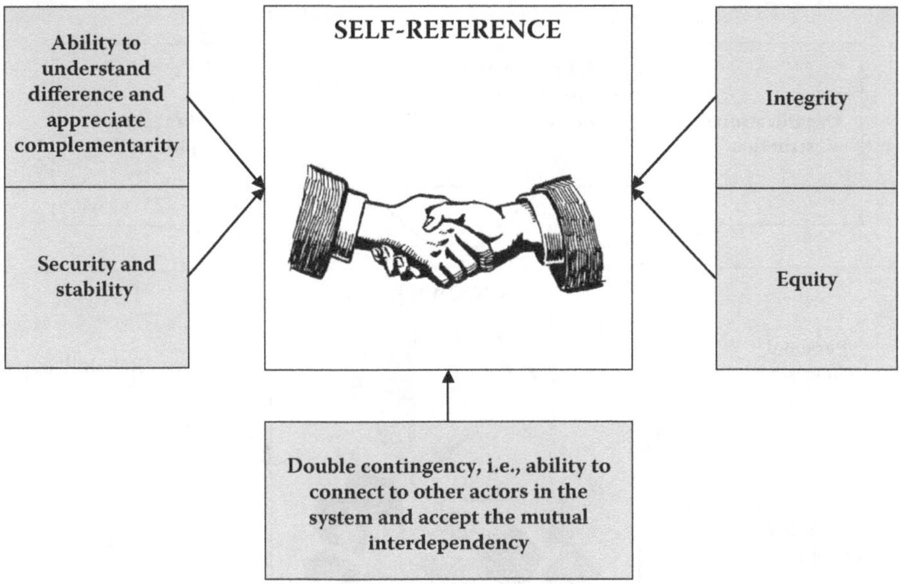

FIGURE 2.4 Self-reference as a component of trust. Source: Adapted from Blomqvist, K., and Pirjo, S. 2004. Trust in technology partnerships. In *Trust in knowledge management systems in organizations*, M.- L. Huotari (author) and M. Iivonen (ed.). Hershey, PA: Idea Group Publishing, p. 189–192.

- **Reciprocity**—expressed values and norms for reciprocity
- **Shared values**—declaration of values, rules of the game
- **Goals and visions**—convergence of goals and shared vision, communication of goals
- **Acceptable code of behavior**—workshops and training, charismatic leadership as an example
- **Personal chemistry**—staff training to understand diversity, informal meetings to increase mutual understanding and socialization
- **Organizational structure**—easy identification of decision-makers and key persons
- **Socialization and creating shared meanings**—shared goals and visions, rituals and symbols, professional norms, shared rhetoric, career and clients
- **Common identification**—informal meetings, interfirm visits, training and education, workshops, group identity
- **Management philosophy**—values and norms.

Behavior is explained by:

- **Shared meanings**—taking care of internal communication, organizational practices and processes to ensure regular communication
- **Multiplexity of communication**—support, advice and emotions included
- **Personal experience**—face-to-face meetings, company visits, product testing, experience of the other company's context

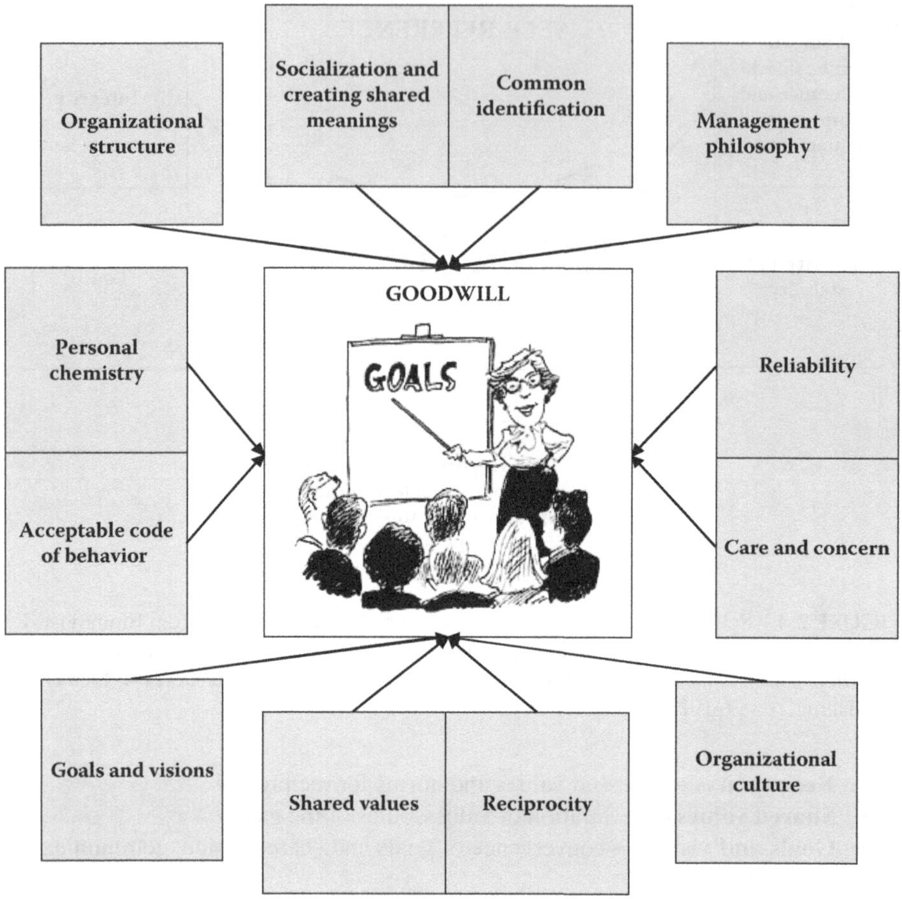

FIGURE 2.5 Good will as a component of trust. Source: Adapted from Blomqvist, K., and Pirjo, S. 2004. Trust in technology partnerships. In *Trust in knowledge management systems in organizations*, M.- L. Huotari (author) and M. Iivonen (ed.). Hershey, PA: Idea Group Publishing, p. 189–192.

- **Learning and understanding**—continuous interaction, interfirm workshops, informal social events
- **Interfirm adaptation**—transfer of key personnel, wide organizational interface
- **Commitment**—attention, expertise, and funds
- **Open commitment**—unilateral investments, incremental investments
- **Shadow of the future**—credible commitments, expectations for future business
- **Information**—sharing sensitive information, sharing future plans related to partner, open and prompt information
- **Communication**—proactive, open and prompt
- **Common base of knowledge**—clearness and frequency of communication.

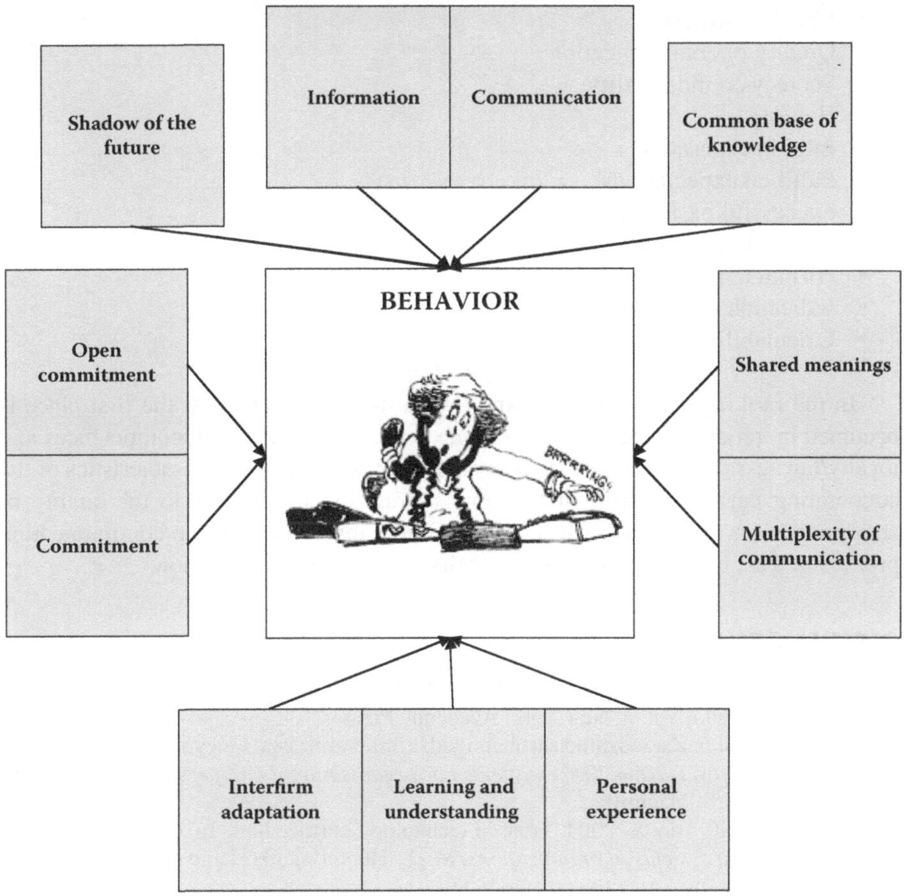

FIGURE 2.6 Behavior as a component of trust. Source: Adapted from Blomqvist, K., and Pirjo, S. 2004. Trust in technology partnerships. In *Trust in knowledge management systems in organizations*, M.- L. Huotari (author) and M. Iivonen (ed.). Hershey, PA: Idea Group Publishing, p. 189–192.

Building trusting relationships is a complex process involving many elements, which were presented and described above in detail. Although these elements have not been explored in detail so far, they have essential implications for the sphere of management. However, each element of the trust-building process has a different sphere of application in practice.

Buessing and Moranz (2003) conducted research on factors that affect the emergence of trust in small-sized enterprises from the sector experienced in Internet-based activities. Factors impacting the development of trusting relations, according to their level of importance (from most significant to least significant), are as follows:

- Reliability
- Professional competences

- Loyalty/fairness
- Quality of communication
- Secrecy/confidentiality
- Honesty
- Intuition/sensation
- Fulfilled expectations
- Successful past projects
- Mode of operation
- Formal agreements/contracts
- Indication of references
- Calculability/predictability.

In the ranking of factors affecting the trust-building process, the first place is occupied by reliability (32% firms), the next ones are professional competences and loyalty/fairness. Reliability and honesty are strongly subjective characteristics of the cooperating partner. Results of the survey draw much attention to the quality of communication. This can suggest the undoubted importance of the communication process in trust building, which shall be discussed in the next section.

REFERENCES

Adams, J. S. 1965. Injustice in social exchange. In *Advances in Experimental Psychology*, L. Berkowitz (ed.), Vol. 2. New York: Academic Press.

Armstrong, M. 2005. Zarządzanie zasobami ludzkimi. Warszawa: Oficyna Ekonomiczna.

Bibb, S., and Kourdi, J. 2004. *Trust matters: For organisational and personal success*. New York: Palgrave Macmillan.

Blomqvist, K., and Pirjo, S. 2004. Trust in technology partnerships. In *Trust in knowledge management systems in organizations*, M.- L. Huotari (author) and M. Iivonen (ed.). p. 189–192, Hershey, PA: Idea Group Publishing.

Buessing, A., and Moranz, C. 2003. Die Rolle von Vertrauen in der Anbahnung telekooperativer Geschäftsbeziehungen. In Telekooperation in Beziehungsnetzwerken für informationsbezogene Dienstleistungen, J. Schlichter, A. Buessing, and R. Reichwald (eds.), 51–84. Technische Universität München (TUM), Institut für Informatik, TUM-INFO-10-I0316-80/1.-FI, October 2003.

Griffin, E. 2003. Podstawy komunikacji społecznej, Gdańsk: Gdańskie Wydawnictwo Psychologiczne.

Koch, M., Möslein, K., Reichwald, R., and Schlichter, J. 2004. TiBiD—Trust Building and Matchmaking Support for Virtual Enterprises, 17th Bled eCommerce Conference eGlobal.

http://www.bledconference.org/proceedings.nsf/Proceedings/D31A5DF7D9590BC1C1256EE000314553/ $File/39Koch.pdf (accessed November 10, 2006).

2.2 INFLUENCE OF COMMUNICATION ON TRUST DEVELOPMENT

The word **communication** comes from the Latin **commercium**, which denotes exchange or connection. This is the way we share information. Before it comes up to any communication, there must be an intention, which takes the form of the transmittable message. This message is passed from the sender (source) to the receiver.

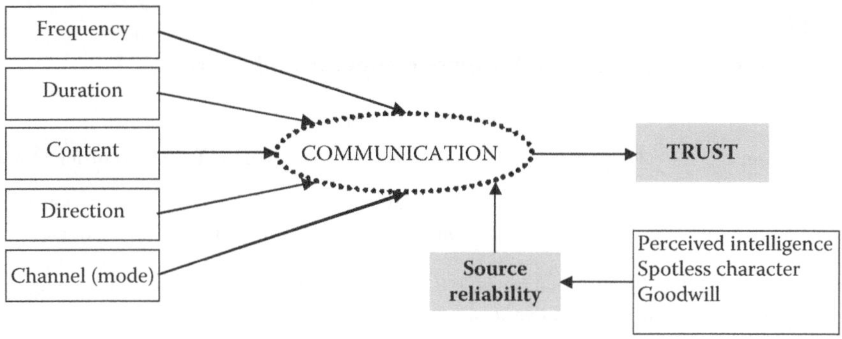

FIGURE 2.7 Influence of communication facets on trust development.

This communiqué is coded, meaning transformed into a symbolic form that is delivered through a channel (mode) to the receiver (Robbins 2004). This receiver, in turn, decodes the announcement initiated by the sender. The communication embraces both the transfer and the understanding of meaning. Inefficient communication is often a source of conflict, whereas effective communication fosters the creation of ties between workers based on mutual understanding, trust, and commitment. It is also a major factor of business relations presented by various models (e.g., Cannon and Perreault 1999, Morgan and Hunt 1994). The model presented by Morgan and Hunt (1994) underlines the affirmative influence of trust on communication. Communication can be simply acknowledged to be the precursor of trust because it refers to the formal and irregular fission (exchange) between actors with essential information, often about strategic character. It is significant in the process of conflict resolution and coordination of representations and expectations. Morgan and Hunt (1994) believe that a high level of trust appears in situations when communication is frequent and of high quality, meaning delivered information was adequate, current, reliable, and understandable (Maltz 2000). Building trusting relationships through communication requires the use of suitable communication strategies, such as frequency, duration, content, channel (of transfer), direction (Figure 2.7).

Frequency and duration of communication refer to quantities of delivered information. **Frequency** defines how often workers communicate with each other. **Duration** of communication is the duration of a single transfer. **Content** relates to delivered information, namely its essentialness, reliabilities, up-to-dateness, and transparentness. Important to content is also the strategy of transfer. The strategy of direct communication is used to change behaviors of other actors by suggestion or request. Examples of such communication are requests, recommendations, promises, and orders. The indirect strategy of communication leads to changes of opinions and attitudes of other workers concerning concrete problems and behaviors when this specific behavior is not directly required. Indirect communication can contain the exchange of information when parties are discussing general business matters, attitudes, and behaviors. The **channel** of transfer determines which method we use for the transmission of information (e.g., telephone, letter, face-to-face conversation) and the degree of formalism. The **direction** of communication for an organization can be horizontal or vertical. Vertical communication is possible in several variants,

TABLE 2.1
Communication Aspects via Business Cooperation Nature

	Communication Aspects			
Cooperation	**Frequency**	**Direction**	**Content**	**Transfer Method**
Relational	High	Two-way	Indirect	Informal
Market	Low	One-way	Direct	Formal

Source: Adapted from Mohr, J., and Nevin, J. R. 1990. Communication strategies in marketing channels: A theoretical perspective. *Journal of Marketing*, 54(4), 42.

such as "top–bottom" and "bottom–top." In interorganizational contacts, the direction of the communication is used for description of levels in organizations which are engaged into the process of the communication between organizations.

The communication in **relational ties** (with long-term orientation and high level of dependence) between business partners is characterized by high frequency, multidirectional information flows, low degree of the formalization, and indirect content of transfer. This builds an affirmative climate of cooperation between partners defined by high levels of trust. Inversely, relations with low levels of trust are characterized by low frequency of communication, one-way information flow, high formalism, and a direct strategy for communication of content. Mohr and Nevin (1990) suggest that the duration of a single communication in conditions of a high level of trust and supporting climate will be short irrespective of the earlier common practice of the communication. The characteristics of communication in view of the relation nature among business partners are seen in Table 2.1.

An information channel is also essential for trust building. Much research has indicated that trust is more difficult to be constituted in relationships using media than in those using face-to-face conversations. The communication grounded on the text is less effective than on relations with high social-emotional content. Recognition of the influence of the medium of communication on trust is very important from the perspective of organizing cooperation at a distance. It is accomplished in experiments called "the social dilemma" (e.g., the game called "the dilemma of the prisoner"). The social dilemma is a situation in which the greatest advantage for the whole group is in conflict with the greatest advantage of every individual, so in consequence both advantages mutually exclude each other. In this situation, if everyone endeavors to the satisfaction of his own needs at the top-level, everyone loses. In many rounds of the game, the maximum advantage for the group appears when every individual agrees to cooperate. Trust level in the group is reflected by full payment from the game. For example, we have four workers who sit in on the four-round game. In every round, each actor has at his or her disposal 200 monetary units that he or she decides to keep or to give back to the group. If they make the decision to invest together with the group, it is possible to obtain a threefold increase of the invested sum; actors experience loss if they decide to keep their own monetary units. The final result depends on trust among group members and is expressed in total revenues of the group from the

TABLE 2.2
Media Classification According to Level and Cooperation Stability

Class	Cooperation Level	Stability Level
Nonverbal communication (identification on the grounds of, for instance, photo, text, references)	Low	Low
Passive communication (prelection, talk show) and audio-conference	Average	Low
"Face-to-face" communication (video conference, "table" conference, direct talk)	High	High

Source: Adapted from Eggert, A. 2001. The role of communication in virtual teams. *Electronic Journal of Organizational Virtualness*, 3(2), 1–7.

investment. Social dilemma games are used for exploration of the influence of communication channels on trust. They can be conducted using different types of media. Various research on the utilization of social dilemmas has demonstrated that direct conversation creates the highest climate of trust. Then preferred in the communication creating trust are properly video conferences, audio conversation, and text. In addition to the level of cooperation, and consequently the level of trust and advantages attained by the group, another important issue is the influence of the medium of the communication on the stability of the cooperation. It comes out that communication via computers is characterized with another course of the trust-building process. The mediation of technological devices forestalls the trust building process, which can be easily observed in communication using only text in the form of e-mail. When there is a lack of gesticulation, facial expression, and subtlety of voice, the decision made by a worker to trust another party and the process of communicating their own reliability can last longer. Furthermore, communication via media is more sensitive to infringements that consequently make the cooperation more fragile than in the case of face-to-face conversations across the enlargement of the probability of opportunistic behaviors. Table 2.2 sums up the influence of the communication medium on cooperation (trust) and the stability of behaviors.

In the process of trust building, there are certain important features of the communication medium, namely: interactivity, assurance of the visibility of communication participants, and exposure of personal information. It is not by fortune that just during direct conversation these three mentioned factors are present as opposed to, for example a photo (only visual identification) or an audio conversation (interaction and revelation of personal information). The greater the resemblance between the course of the communication processes with the means of the concrete medium to the direct conversation, the greater its ability to build trusting relationships.

Electronic communication can turn out to be inadequate for an organization that does not have prior experiences from mutual cooperation. Trust in this situation can be initiated by initially enabling the members to have a direct conversation. When building trusting relations in the process of the communication: the following are essential:

- Portioning confidential information
- Communicating future plans concerning the partner
- Providing open and immediate transfer of information (opinions)
- Exposing affirmative and negative aspects of the cooperation
- Organizational practices and processes assuring regular communication.

In addition to the essential elements of trust building through communication, equally important is the avoidance of activities that destroy trust, as trust is inherently very fragile and vulnerable to single incidents. Because trust building is a long-lasting phenomenon, its undermining is very impetuous. Therefore, very meaningful input was made by Covello (1992), who prepared a list of pitfalls that are said to be the potential sources of problems in communication processes. The most essential pitfalls from the view point of trust management are described below, simultaneously with suggestions for how to decrease their negative influence on trust (what to do and what not to do):

1. **Jargon**—We unquestionably have to define any technical and professional terms that are not known to the general public. We should not use language that anyone in our audience does not understand.
2. **Humor**—Humor is best when it is pointed directly at us. A good sense of humor, especially the ability to laugh at our own behaviors, weaknesses, and so on, is well seen. It means that we are able to distance ourselves from what had happened and overcome hurdles in any circumstances. We should not speak humorously in relation to environment, health, and safety issues.
3. **Negative allegations**—The best solution is to refute negative allegations, rebut any negative statements without repeating them. It is not a good idea to refer to them or come back to them in any way.
4. **Negative words or phrases**—Terms with positive or neutral character are the most suitable in every situation. We definitely should avoid any issues that are the subject of much controversy.
5. **Reliance on words**—To pass the pivot and highlight the major points, we should use accessible visuals. We should never limit ourselves to only pure words. Words are just one means and, nowadays, there is a huge variety of other means using existing technology.
6. **Temper**—We ought to stay calm and quiet. To make a better impression, we might use a question or allegation from our audience as a springboard to say something nice and positive. We should control our feelings and balance them because too much passion hinders clear, smooth communication.
7. **Clarity**—We are obliged to check whether our audience understands us and whether we speak clearly enough. To do so, we should ask for questions or, even better, ask the audience questions about what was explained so far. Moreover, we ought to look for confirmation from the audience that we can continue and the communication is flowing correctly. A very big mistake is assuming that everything that was said is completely understood.

8. **Abstractions**—In order to raise the level of familiarity of our audience with the topic, we can use some examples, stories, and analogies tailored to the audience. This technique can enhance the general understanding.
9. **Nonverbal messages**—We cannot forget about body language, nonverbal signs that are being sent to the audience. We have to be aware of them and create them to enhance what we want to communicate, to be coherent and consequent. However, we should avoid being too uptight and artificial. Sometimes, it is very helpful to learn your body language, to listen to opinions about one's way of communicating, or even to record oneself and make conclusions from it for the future.
10. **Attacks**—One is allowed to attack a problem or issue, but at the same time it is forbidden to make personal attacks or accuse the organization.
11. **Promises**—We have to be careful about making promises to be sure that they are reliable and attainable, that we can succeed in delivering them. It is not acceptable to "promise the moon" or not to keep promises made.
12. **Guarantees**—We should accent successes, gained targets, made accomplishments, and efforts in the process. We should know there are no guarantees.
13. **Speculation**—We need to be as concise and concrete as possible, delivering information on facts, what was or is being done. We should not discuss the worst cases because they provide only negative, pessimistic feelings amongst the audience.
14. **Organizational identity**—It is advised to use personal pronouns (I, we) to be recognized as an individual. We ought not adopt the organizational identity of a huge organization, which is very often perceived as impersonal, formal, and stiff.
15. **Blame**—We cannot escape from taking responsibility for our participation in the problem. Furthermore, we cannot transfer blame or responsibility to others.
16. **Numbers** Words need concrete proof, especially numbers. Research results can help us explain ongoing or past trends, gained growth, and decrease or stagnation. However, we have to avoid presenting enormous negative numbers because they can leave an unwanted final impression about the topic, influence our overall communication process badly ruining what was gained so far by us.

We ought also to remember that the effectiveness of communication is mainly affected by the **reliability of the source.** Before now, Aristotle (cited in Griffin, 2003, p. 309-310) in "Rhetoric" described three features that qualify the reliability of a source:

1. **Perceived intelligence.** Intelligence ties in with the practical wisdom and, in tandem, believed values. Receivers judge intelligence on the grounds of the relation occurring between their own opinions and the ideas of the orator.
2. **Spotless character.** Character clearly influences an orator's image as a good and honest man.

3. **Good will.** Good will stands for the orator's affirmative intention to his receivers. Aristotle reckoned that we could be extremely intelligent and have character but not take into consideration the receivers' well-being.

The reliability of the source is also influenced by the context of the communication: the acquaintance of the sender, the kind of ties between the sender and the receiver, and so forth.

REFERENCES

Bos, N., Olson, J. S., Gergle, D., Olson, G. M., and Wright, Z. 2002. Effects of four computer-mediated communication channels on trust development. In *Proceedings of Human Factors in Computing Systems* (CHI 2002), 135–140. New York: ACM Press.

Cannon, J. P., and Perreault, W. D., Jr. 1999. Buyer-seller relationship in business markets. *Journal of Marketing Research*, 439–460.

Covello, V. T. 1992. Risk communication, trust and credibility. *Health and Environmental Digest*, 6(1), 1–4.

Eggert, A. 2001. The role of communication in virtual teams. *Electronic Journal of Organizational Virtualness*, 3(2), 7.

Griffin, E. 2003. Podstawy komunikacji społecznej. Gdańsk: Gdańskie Wydawnictwo Psychologiczne.

Maltz, E. 2000. Is all communication created equal?: An investigation into the effects of communication mode on perceived information quality. *Journal of Product Innovation Management*, 17, 110–127.

Mohr, J., and Nevin, J. R. 1990. Communication strategies in marketing channels: A theoretical perspective. *Journal of Marketing*, 54(4), 35–51.

Morgan, R. M., and Hunt, S. D. 1994. The commitment trust theory of relationship marketing. Journal of Marketing, 58, 20–38.

Robbins, S. P. 2004. Komunikacja. In Zachowania organizacyjne. Warszawa: Polskie Wydawnictwo Ekonomiczne.

2.3 MANAGING TRUST IN LEADERS

Trust is a key element of effective leadership (Likert 1967, McGregor 1967). Trust in a leader bears on results of the subordinates' work, job satisfaction, organizational engagement, and commitment to realization of the of the leader's decisions (Dirks and Ferrin 2000). Simons, McLean, and Parks (2002) investigated how a firm is influenced by an honest behavior of managers creating trust (trust to the managers). Based on a sample of hotels, they pointed out that perceived honesty of managers' behaviors, expressed in them by workers' trust, was positively connected with customer satisfaction and profitability of the activity. They presented a practical model that explained as much as 13% variation in profitability, where trust played a key part in this effect. Trust in a leader is also connected with performance of workers' groups and teams, as discovered by Dirks (2002) in research covering 30 National Collegiate Athletic Association basketball teams. Considering different determinants of teams' results (e.g., talent of players, cadences, experience and achievements of the coach, results in previous seasons, results in previous years, trust among team members), research displayed that trust in the coach (the team leader) explained 7%

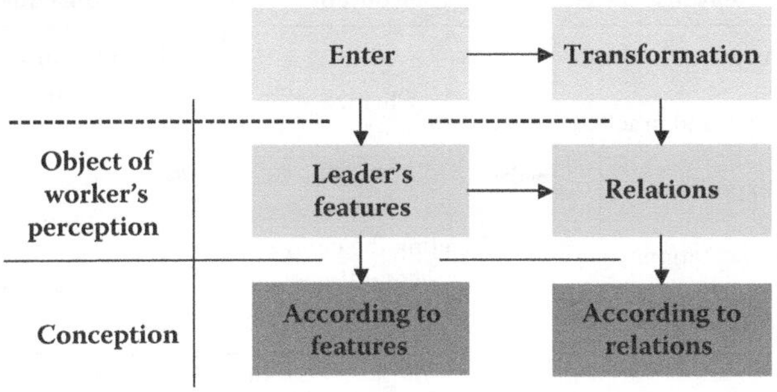

FIGURE 2.8 Conceptions for leader's trust perceived by workers.

of variations in teams' results. What is more, contrary to expectations that trust in remaining team members is a key for success, it turned out that team results could be better predicted by trusting the team leader rather than by trusting team members. Furthermore, trust in leaders was smaller than expected. These results probably come from the fact that trust is very sensitive to single incidents. Every single negative event affects trust, causing its decrease and making the supporters more sensitive to future disturbances.

Recognition of the key aspects connected with trust in the context of leadership is fundamental for the organization management. However, multiple disciplines (e.g., organizational psychology, organizational communication, education, public administration) have been dealing with the issue of trust management among leaders for the last decades. In this section, we accept the simplification that under the name "leader" we understand both the leader and the manager. In management practice and literature, there is an accepted general opinion that a leader is simply a good manager (Kożusznik 2005, p.102). We assume for the purpose of this book that we do not see the distinction between leader and manager. In the literature concerning trust in the context of the leadership, we distinguish fundamentally two approaches (Figure 2.8).

The first approach is perspective founded on relations and the second on character (Dirks and Ferrin 2002). **Trust perspective based on features** relates to perceived character of the leader by subordinates, which is important because the leader can have authorization to undertake essential decisions concerning results experienced by subordinates (e.g., promotions, lay-offs, and task assignments). In this approach, we commonly formulate a list of features referring to the leader that determine the leader's reliability. If workers are convinced about the dishonesty of the leader, they will unveil the concentration on the avoidance of results of this fact and, consequently, they will lower the effectiveness of their own work. **Trust perspective based on relations** concentrates on the nature of the relation between the leader and supporter and, speaking precisely, how the supporter perceives the nature of the relation and the behaviors of the leader. Central facets of high-quality relations are care and consideration. Trust is an element of the psychological contract between parties (Rogers 1995, p.15)—that is to say, the hidden contract between individuals that specifies what everyone receives and gives in mutual interactions. A supplement of the formal contract is, for instance, service contracts. The psychological

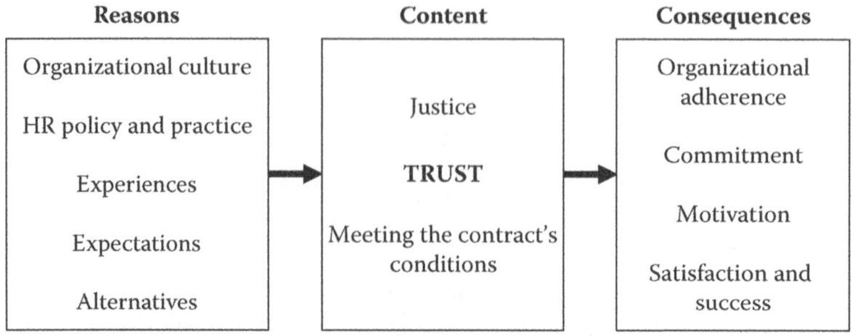

FIGURE 2.9 Psychological contract model. Source: Adapted from Armstrong, M. 2005. Zarządzanie zasobami ludzkimi. Warszawa: Oficyna Ekonomiczna, p. 281.

contract derives from unwritten, sometimes inarticulated contracts between parties, coming to light in the system of the conviction about mutual obligations. These obligations have come into being since the day concluding the formal contract, across the process of employment, continuous in everyday interactions, and have their own tangible signs as promises and intangible ones such as tone of voice and body language. The operating model of the psychological contract proposed by Guest (cited in Armstrong 2005, p. 281; Figure 2.9) erects elements on which the contract is founded—justice and trust, and the degree to which the transaction is expressed directly—is perceived as being realized in the sense of keeping of promises.

Contributions to the understanding of trust in the context of leadership can help develop the science of **neuroeconomy** (Żak 2004, p. 1737), which seeks in neurobiological sources of economic decisions answers to the question: "Why do two persons having the same amount of information make different decisions?" It investigates among other things the influence of oxytocin (Kosfeld et al. 2005), mimicries, and the body language of other people and analyzes the influence of facial expression on trust (Winston et al. 2002). We can acknowledge **neuroeconomy** as the physiological trend explaining trust. The feature of being reliable is derived from many different aspects: aromatic sensations, sounds affecting activity of the amygdaloidal body responsible for emotional and motivational processes and influencing the estimation of the second person without a will (the intuitive experience) if the given person is reliable. Neuroeconomical research (in the form of the economic game) character proved that oxytocin, a neurohormone, is responsible for trust among people (Kosfeld et al. 2005).

We ought to remember that the distinction between these two perspectives is not so obvious in real situations because these perspectives mutually intermingle. Concrete character traits result in concrete behaviors in relations. The issue is the systematization of knowledge and the possible measurement of trust in managers. Perspectives based on features are founded mostly in character, which is constituted by temperament, elements of individuality. "The temperament describes mostly physiological factors of the lower row and is strongly determined genetically, while the character refers to higher cognitive activities, as well as the attitude and values which affect learning processes and the environment" (Hoffman 2004, p. 32). The approach to trust resting on features and, more exactly, on workers' perception of the

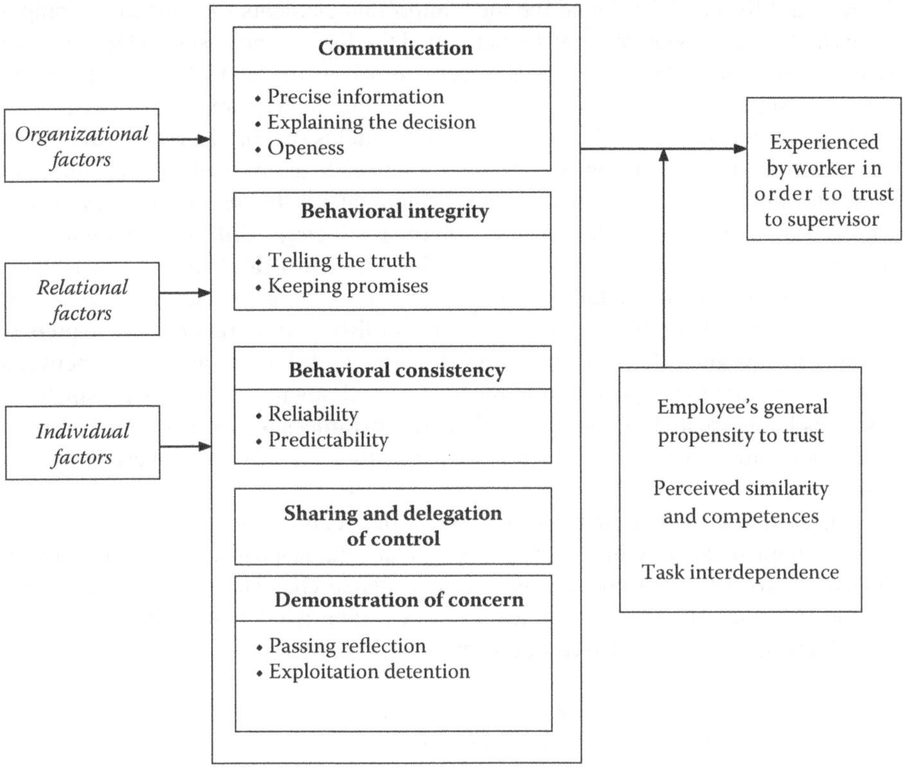

FIGURE 2.10 Exchange framework of initiating managerial trustworthy behaviors according to Whitener and colleagues. Source: Adapted from Whitener, E. M., Brodt, S. E., Korsgaard, M. A., and Werner, J. M. 1998. Managers as initiators of trust: An exchange relationship framework for understanding managerial trustworthy behavior. *Academy of Management*, 23(3), p. 519.

character of the leader does not exclude the physiological factors of the lower row presented neuroeconomy.

The approach based on features is presented, for example, by Mayer et al. (1995), Butler (1991), and Kee and Knox (1970). The perspective based on features of the leader refers to perceptions by the workers and not objective proprieties of the relation or the leader himself. In this light, trust—which the worker bestows on a superior—is measured. Mayer et al. (1995, p. 717–720) identified three qualities of a superior having the influence on trust experienced by workers: 1) ability—skills and competences the partner (manager) possesses with respect to the specific domain, 2) benevolence—whether the partner functioning at changing circumstances works in favor of the second party, and 3) integrity—whether the partner keeps written and verbal promises. Of course, these are necessary elements, but not sufficient, so that a worker trusts an employer (Figure 2.10). Always there acts the general propensity to trust of the particular worker (for more detail, see the section "Factors Affecting Generalized Trust").

Kee and Konx (1970) found the most important elements for trust to be competences and motives. Gabarro (1978) elaborated the list to nine bases (antecedents) of trust, namely: **integrity** (the leader is sincere), **motives** (the leader has friendly intentions), **consequence** (the leader behaves predictably and is unfaultable), **openness** (the leader freely expresses his or her own ideas and is open), **discretion** (the leader keeps secrets), **functional or specific competences** (trust in skills, competences of the leader in the realization of concrete assignments), **interpersonal competences** (trust in social skills of the leader [e.g., solving conflicts]) , **skill of judgment** (trust in fairness of decision made by the leader), **"business intuition"** (trust in the leader's common sense and understanding of business activity rules). Butler (1991) introduced a list of 10 conditions for trust: **accessibility, competences, consequence, discretion, integrity, loyalty, openness** (frankness), **fairness** (agreement between the treatment and the expectation of the worker with regard to the way in which the worker is being treated by the leader), **keeping promises**, **openness** (contradistinction to dogmatism, openness to new ideas). The three last elements were not previously identified by authors.

There exists a relationship between the individuality of the manager and managerial practices implemented by him. Concrete individual characteristics are connected with applied control style and consequently these concrete characteristics can raise trust. Kramer and Tyler suggest that "five features" are responsible for creating trust, adapted from **Briggs-Myers Trust Inventory,** namely (cited in Martins 2002, p. 758–759):

- Conscientiousness—Embraces such features as being organized and hard-working and also being responsible, trustworthy, and unfailing.
- Agreeableness—Reflects cheerful nature, collectivity, forgiving, and amicability. The opposite pole would be cold, rude, unkind, and independent.
- Emotional stability—Reflects the lack of tense, depression, anger, worries, and sense of trouble. The opposite pole is called neuroticism.
- Openness to experiences—Refers to creativeness, imaginativeness, intelligence, and broad-mindedness.
- Extraversion—Determined by sociability, cheerfulness, talkativeness, and activeness.

The influence of the model "of five features" affecting trustworthiness is still not verified. Nevertheless, this concept can help to determine a tool that is able to support the process of candidate selection for managerial positions. Thus, one can compare a candidate's profile with the standardized profile of the individuality of the manager who transmits behaviors, creating an atmosphere of trust between him and his subordinates.

The approach to trust based on relations was presented by Whitener et al. (1998), who elaborated the model illustrating junior subordinates' trust in managers. They qualified managers' behaviors that have an influence on trust in them. They defined managerial trustworthy behaviors as "volitional actions and interactions performed by managers which are necessary, though not sufficient to engender employees' trust in them" (Whitener et al. 1998, p. 516). They proposed a taxonomy of five categories of managerial behaviors that influence employees' perceptions of managerial trustworthiness, founded on previous research on the antecedents of trust (e.g., Butler 1991). These categories actually apply to features and, therefore, we can clearly see the optionality of the division into the perspective of trust in a leader based on features and based on relations. Five behaviors affecting an employee's experienced trust in a leader are as follows:

1. Behavioral consistency in mutual interactions
   - Reliability
   - Predictability
2. Behavioral integrity
   - Telling the truth
   - Keeping promises
3. Sharing and delegating authority
4. Communication
   - Information accuracy
   - Explanation of decision
   - Openness
5. Demonstration of concern
   - Showing consideration
   - Forbearance versus taking advantage

Dirks (2006) makes some interesting remarks about the relationship of dependence between integrity and trust. He believed that people with high integrity would restrain themselves from dishonest behavior irrespective of the situation, whereas people with low integrity might behave honest or dishonest depending on the specific situation and their opportunities, goals, and incentives. As a consequence, a single honest act is recognized as a signal of honesty in certain situations. This applies to both groups, with low and high integrity. However, what is more surprising, a single dishonest act is interpreted as a signal of low integrity and it is assumed that only a person with low integrity would behave dishonestly.

Whitener et al. (1998) also elucidated factors other than the manager's behavior that affect employees' trust, most notably organizational factors, relation-based factors, and individual factors. To organizational factors belong **organizational structure, corporate culture,** and **human resources (HR) management practices.** The choice of the organizational structure influences the dimension of organizational control. The network organization entails the necessity for behaviors creating trusting relationships between members. Organizations with a large emphasis on control that are highly centralized and formalized constitute a rather unfriendly environment for the trust atmosphere to be created. Organizations with low control that are highly decentralized and low-formalized foster managerial trustworthy behavior—in particular, delegation of control and open communication of ideas. Applied organizational HR policies and procedures (e.g., incorporation of procedural justice principles in performance appraisal and reward systems, regular and timely feedback on performance estimations, the right to abrogate the results of a performance appraisal, well-defined and transparent criteria of assessment) are likely to stimulate managerial trustworthy behaviors, such as behavioral consistency and appropriate communication. The culture also influences the managerial behaviors in the organization. Organizational cultures that value risk-taking behaviors are likely to support managers who delegate control to subordinates regardless of the outcomes. Cultures that declare interpersonal values, advocate open communication, and are human-being oriented reward managers for collaborating, sharing information, explaining decisions, openly discussing problems, and expressing the care of their own workers.

Relational factors refer to initial **interactions, expectations,** and **costs of the exchanges.** The initial interaction relates to establishment of a very strong relationship based on the manager's first affirmative impression on new workers' abilities and his or her response to the expectations of the workplace. The more favorable initial interactions are, the more probable it is that a manager will delegate control to an employee, openly communicate information, and show concern, which in turn are likely to enhance the employee's trust in the manager. If a manager initially perceives that a worker meets the requirements, responds to the exchange, and shows his reciprocity, then the higher the likelihood is that more managerial trustworthy behaviors will occur, such as power delegation or frankness in communication (the primacy effect and so-called self-fulfilling prophecy). What is not surprising is that the costs of exchanges are considered, too. They are connected with opportunistic behaviors of the worker. For example, control delegation in very important projects can have substantial negative repercussions for the manager if a worker shows a lack of loyalty. This can hold back enactment of trust, such as sharing strategic information, because delegation allows employees with greater freedom to act opportunistically.

Individual factors pertain to the general propensity to trust, self-efficiency, and values shared by individuals. Managers with a **greater propensity to trust** (with a higher level of general trust) more often will perceive their own workers as persons that potentially reciprocate trust and, consequently, will be more engaged in trustworthy behaviors. The **self-efficiency** of the manager refers to his knowledge, skills, opportunities, and abilities. For instance, a manager with low self-efficiency concerning his or her ability to control delegation will be reluctant to engage in practices of participative management processes. Similarly, in the case of the lack of managerial skills regarding conflict management, the manager can avoid open, multidimensional communication. **Values** of the manager consist of goals, their importance, motives, attitudes, and choices. Worth mention is the distinction between self-transcendence, expressed in universalism and benevolence, and self-enhancement, invoking Hedonism, Machiavellianism, and Particularism. Managers with universalist values (understanding, appreciation, and protection of the welfare of all people) are likely to demonstrate concern for workers more often than managers with a strong need for power and dominance over other people. Managers who value benevolence (enhancement of the welfare of others) more often keep promises, tell the truth, and behave honestly than managers who declares hedonism (self-gratification). Values held by the manager determine "the internal compass" promoting definite dimensions of trustworthy behaviors, including demonstrating concern, behaving consistently, and behaving with integrity. This compass is, of course, modified by organizational factors and relational factors, which are transmitted as a whole to definite managerial trustworthy behaviors.

On trust experienced to the leader apart from elements quoted by each authors also has influence a period of the mutual interaction. Trust in a leader affects behaviors and attitudes of employees and their performances. Trust plays a critical part in crisis (e.g., dismissals, cuts in earnings). Trust in these periods can be enlarged by ensuring qualitative communication (e.g., face to face), transferring an adequate decision explanation, and treating employees with dignity and respect. Folger and Skarlicki (1998) drew attention to sore points that managers are likely to avoid in interpersonal contact with workers in cases when there is a necessity to convey bad news, especially, when they are convinced that they are somehow responsible. This paradoxically explains why managers tend to avoid direct contacts and behaviors enacting trust as well as helping to cope with the crisis. As a matter of fact, this is a very good illustration of the importance of understanding the previously described dimensions responsible for an employee's perceived trust in a leader.

Given that trust in leaders does facilitate essential outcomes for organizations, a focal question is: How is trust management built and rebuilt in an organization and what factors undermine trust in leaders? The idea that the behavior of managers influences trust is common, but it is too general and needs to be clarified, explained in detail. The conventional wisdom about trust in leaders is that trust takes a long time to build but can be ruined in a second. Therefore, we require an answer to the following question: What types of behaviors are a good (or bad) foundation on which to built trust in leaders? Trust management for managers implies the avoidance of all situations that can destroy it. The recovery of abused trust of employees in management by nature is difficult or even impossible; still, we might undertake attempts at its recuperation. Herriot, Hrish, and Riley (cited in Armstrong 2005, p. 274–277) postulated the trust recovery four-gradual program consisting of the following issues:

1. Top management is not sufficiently devoted to various needs of workers.
2. Taking into account the individual needs of workers and their opinions about the new work methods during their implementation.
3. Building trust based on knowledge results not from the definite transaction but from the fact that the organization wants to be perceived as reliable and not as treating trust instrumentally.
4. Building trust grounded on the acknowledgement that every party shows empathy to the second party's needs (although in practice this state is difficult to achieve).

Considering the fact that communication affects the enactment of trust, we ought to take care also for its efficiency. Primarily, this is a fundamental issue in the communication process of unwished messages (e.g., in manager–subordinate relations). The keystone here is to use **feedback** to avoid misunderstandings and irregularities in the communication process. The advice listed below can be very helpful for managers in enlarging the efficiency of delivered feedback:

1. Focus attention on hard facts. We ought to avoid general statements. Feedback should concentrate on hard, pure facts.
2. Feedback should refer to behaviors and not to a person's character traits. We ought not criticize persons, but their behaviors. Telling people that they are "stupid" or "incompetent" almost always brings results opposite to the intentions.
3. Feedback should refer to targets. If a manager has to say something negative, he should ensure that it concerns the aims of the receiver and the scope of his empowerment. The manager should know who needs the feedback. If he comes to the conclusion that he simply wants "to vent to his feelings," then he should not speak about this.
4. Choose a suitable moment for information conveying. Feedback has the greatest meaning for the addressee if there is a very short period between the behavior and receiving feedback about this behavior. We ought to remember the effect of freshness in this case.
5. Assure transparency. Is the feedback rich enough in content and exhaustive so that the addressee understands the message clearly and fully? We can think out, whether it is proper to ask receivers for paraphrasing the transferred content, to make sure the intended meaning was properly understood.
6. Negative feedback should refer to behaviors that depend on the receiver. There is no point in reminding somebody about inadequacies on which he has no influence at all. Negative feedback especially should refer to behaviors about which the receiver can do something (Robbins 2004, p. 219–245).

During conversations with subordinates, we ought to remember also about active listening to the worker; to understand contents delivered by the worker. The following advice can help managers listen attentively:

1. Ensure eye contact. Ensuring eye contact with the speaker concentrates the attention on him and limits the propensity to distract him. Finally, it encourages the speaker to state his opinions.
2. Giving a nod and carrying the suitable facial expression. An effective listener shows interest in the speaker by means of nonverbal signals.
3. Avoid distracting acts or movements, for example, looking at one's watch, throwing away papers, or playing with a pen. These behaviors suggest that the listener is bored and uninterested in what is being said.
4. Ask questions. Questions lead to explaining doubts and ensuring mutual understanding.
5. Paraphrase. Paraphrasing enables one to check whether somebody is attentively listening and whether the listener understood what was said.
6. Avoid interrupting. Let the speaker finish his own thoughts before you answer or add something. Do not try to guess what the speaker is trying to say.

The intent of this section was to draw attention to and explore some aspects of trust in leaders. It is crucial to understand trust in leadership. Contemporaneously, we want to stress the fact that a lot of questions still have not been answered and tested in this area. Summing up the considerations, we might create some guidelines for managers who want to build reciprocal trust.

Be open. Cynicism descends to the same degree from this what people do not know and from what they already know. Openness leads to trust and reliability, so inform your workers, lead to the full clearness of decision-making criteria, explain the logic of your decisions, and sincerely introduce problems.

Be fair. Before making a decision or undertaking an activity, think out how it will be perceived by others from an objective point of view. Give others the possibility to make a contribution, be objective and impartial in the estimation of the effectivity, and pay attention to the sense of justice in allocation of rewards. Remember existing theories of justice (e.g., Smith's theory).

Tell the truth. Uprightness has a key meaning in trust building, so you must be perceived as a truth-telling person. In general, workers would prefer to hear something unwanted than to be lied to by their managers.

Demonstrate the consequence. Workers expect predictability. Comfortably think over your own values and beliefs. Then consistently use them in the decision-making process.

Keep promises. For trust, it is indispensable that people are convinced that they can rely on you. Keep promises and meet your obligations.

Keep secrets entrusted to you. People trust other people who are discreet and on whom they can rely. If workers are open to you or tell you a secret, it is necessary to them to have the certainty that you will not talk it over with other people or betray their secrets.

Demonstrate competence. Gain admiration and respect from others by demonstrating your technical expertise. Pay special attention to acquiring and showing social competences, such as leadership and communication.

Despite the rising awareness of the high importance of trust in leaders, still many issues stay unsolved. Knowledge about building and maintaining trust in leadership is expanding. The reason behind the failure to manage trust in leaders is constant: leaders are focused on other goals that seem to be more essential for them and their organization; trust is in a lower position in their hierarchy of values. Therefore, leaders need to understand and remember at all times that trust is undoubtedly an initial condition to build leadership and there is no successful leader without trust management.

REFERENCES

Armstrong, M. 2005. Zarządzanie zasobami ludzkimi. Warszawa: Oficyna Ekonomiczna.

Butler, J. K., Jr. 1991. Toward understanding and measuring conditions of trust. *Journal of Management*, 17(3), 643–663.

Cardona, P., and Elola, A. 2003. Trust in management: The effect of managerial trustworthy behavior and reciprocity. WP, No. 496, University of Navarra, Spain.

Dirks, K. T. 2000. Trust in leadership and team performance: Evidence from NCAA Basketball. *Journal of Applied Psychology*, 85(6), 1004–1012.

Dirks, K. T. 2006. Three fundamental questions regarding trust in leaders. In *Handbook of trust research*, R. Bachmann and A. Zaheer (eds.), 15–28. Cheltenham, UK: Edward Elgar Publishing, Ltd.
Dirks, K. T., and Ferrin, D. L. 2002. Trust in leadership: Meta-analytic findings and implications for research and practice. *Journal of Applied Psychology*, 87, 611–628.
Folger, R., and Skarlicki, D. P. 1998. When tough times make tough bosses: Managerial distancing as a function of layoff blame. *Academy of Management Journal*, 41, 9–87.
Gabarro, J. J. 1978. The development of trust influence and expectations. In *Interpersonal behavior: Communication and understanding in relationship*, J. J. Gabarro (ed.), 290–303. Englewood Cliffs, NJ: Prentice-Hall.
Hoffman, E. 2004. Ocena psychologiczna pracowników., Gdańsk: Gdańskie Wydawnictwo Psychologiczne.
Kee, H. W., and Knox, R. E. 1970. Conceptual and methodological consideration in the study of trust. Journal of Conflict Resolution, 14, 357–366.
Kosfeld, M., Heinrichs, M., Zak, P. J., Fischbacher, U., and Fehr, E. 2005. Oxytocin increases trust in humans. *Nature Neuroscience*, 435(2), 673–676.
Kożusznik, B. (2005). Wpływ przywódcy w organizacji. In Wpływ społeczny w organizacji, 102–121. Warszawa: Polskie Wydawnictwo Ekonomiczne.
Likert, R. 1967. *Human organization: Its management and value*. New York: McGraw-Hill.
Martins, N. 2002. A model for managing trust. *International Journal of Manpower*, 23(8), 754–769.
Mayer, R. C., Davis, J. H., and Schoorman, F. D. 1995. An integrative model of organizational trust. *Academy of Management Review*, 20, OH, 707–734.
McGregor, D. 1967. The professional manager. New York: McGraw-Hill.
Robbins, S. P. 2004. Zachowania organizacyjne, Warszawa: Polskie Wydawnictwo Ekonomiczne.
Rogers, R. W. 1995. The psychological contract of trust: Part I. *Executive Development*, 8(2), 15–19.
Simons, T., and Parks, J. M. 2001. Empty words: The impact of perceived managerial integrity on employees, customers and profits. Working paper.
Whitener, E. M., Brodt, S. E., Korsgaard, M. A., and Werner, J. M. 1998. Managers as initiators of trust: An exchange relationship framework for understanding managerial trustworthy behavior. *Academy of Management Review*, 23(3), 513–530.
Winston, J. S., Strange, B. A., O'Doherty, INITIALS, and Dolan, R. J. 2002. Automatic and intentional brain responses during evaluation of trustworthiness of faces. *Nature Neuroscience*, 5(3), 277–283.
Żak, P. J. 2004. Neuroeconomics. *Philosophical Transactions of the Royal Society* B, 359, 1737–1748.

2.4 TRUST BUILDING IN RELATIONS BETWEEN BUSINESS AND CLIENT (B2C)

Trust in somebody or something denominates that the other side shares our norms and values and will be acting to get benefits for us and will not cause us any harm (Lewicka-Strzałecka 2003, p.197). Then, the stronger our conviction is about this, the higher is our readiness for cooperation and consequently grow chances that this cooperation will end with success. Trust has a true economic value. In this section, we draw attention to trust building in relations between enterprise and customer. Because the pivotal theme of this book is trust management in virtual organizations (VOs), below are characterized mainly relations in the electronic economy in B2C relations.

We can confide not only in a man but also, for example, in an enterprise, institution, or organization constructed by the man. Trust is a specific subjective probability. In addition, trust can be expressed in certain degrees. The trust degree is a decisive factor conditioning the decision-making process in enterprise-customer relationships (Lewicka-Strzałecka 2003, p. 197).

Consumer trust is trust of a customer in the enterprise, which is considered to be a special kind of trust. The consumer is a partner in the exchange act that takes place on the market. The relationship customer-enterprise is characterized by a strong asymmetry. There appears an individual (or set of individuals) and the organization (the enterprise). The organization enjoys not only financial and technical advantages but mainly advantages based on information. In this situation, consumer trust stands as the subjective conviction of the customer that the enterprise will not use its own superiority against him.

Consumer trust in e-commerce is recognized as customer frankness concerning activities of an Internet vendor based on the expectation that the salesman will act in a generally accepted manner, regardless of this, whether the customer will use the possibility of checking and controlling this Internet vendor (Lee and Turban 2001). In general, this specific kind of trust is called **online trust.** Confidence is based on experience and emotions and is accompanied by uncertainty, risk, dependence of the customer on Internet vendor, and the possibility of vendor opportunistic behaviors (Lewis and Weigert 1985, Moorman et al. 1992, Uzzi 1999). **Off-line trust** refers to traditional trade, where the Internet is not applied to gain new potential customers.

In his book *Trust: The Social Virtues and the Creation of Prosperity,* Francis Fukuyama (1995) argues that a strategic factor affecting the development of the sector of so-called "new economy" is the degree of mutual trust within the society. In the present, developed world, more and more transactions are conducted distantly. Information is becoming the main value. Properly high level of trust is a necessary prerequisite required to gain access to the personal data of a consumer who is aiming at making online shopping, without first-hand contact with the commodity or the salesman (Kossecki n.d.).

In the customer-enterprise relation, one can distinguish three modifications of trust: **calculus, institutional, and personal.** One of the main elements of consumer trust is the calculation. Calculus consumer trust is based on the information passed, for example, on etiquettes, guarantees, tests of quality, and findings. Its source is the past own and social experience with the definite enterprise. Institutional trust is founded on existing legal regulations and activities of consumer organizations. Therefore, a major part is played by the institutional brand, which is a guarantee of the constancy to definite parameters. Additionally, apart from the well-known and certain brand-name value of products, some customers purchase products for the prestige or owing to the current fashion. In personal trust, the customer entirely believes in the given brand, its controls, and considers all undesirable situations as something very rare and totally coincidental. Thus, customers for some time may somehow not notice disadvantageous information on assessed products and do not take this type of information into account during purchase (Hoyer and Brown 1990).

After deliberations over the subject of trust relations between the enterprise and the customer, we can point out three kinds of consumer risk. **Functional risk** connotes uncertainty about whether the product will perform its function or the producer will keep all contract stipulations, particularly those connected with the service. **Financial risk** is expressed in uncertainty regarding the suitability of the product price and hidden costs that need to be covered in the future. The **risk of undesirable side effects** refers to possible unwanted and unpredicted side effects connected with the exploitation of the product.

The existence of consumer trust permits the diminishment of the complexity accompanying the decision-making process on the market. The customer is not able to acquire all information about the specific product. Trust signifies customer faith that the enterprise will not take advantage of him in opportunistic actions. Substantial here is the emotional factor of personal trust.

Building trusting relations between the enterprise and the customer is performed using the **persuasive announcement,** which targets the change of attitudes (Caccioppo and Petty 1981). This kind of announcement might be delivered through two channels: central and peripheral. Information delivered through a central channel is very attentively accepted and converted, and there is needed a lot of time and many definite skills and certain knowledge. In turn, the information derived from a peripheral channel is taken automatically, without any reflection or additional effort. Some reckon that it is easier to acquire customer trust by peripheral channels. However, here still exists the risk that the announcement will later undergo to rational estimation and perhaps it will be rejected. For example, a promise by an academy of health that you will lose weight four times quicker might be at the very beginning accepted rapturously. Later on, it becomes obvious that this promise never will be realized.

Customer trust level in enterprise is determined by a broad range of factors to which belong, among other things, disposable funds of the customer, type of purchased good, and type of competition on the market for the certain good (Lewicka-Strzałecka 2003, p. 201). When a monopoly exists in a certain market, the customer is not given the freedom of choice. The customer's decision can proceed double-track: to embrace the offer (to trust) or not accede to the transaction (to distrust). Enterprises endeavoring to gain the competitive advantage should build trustful relationships with their customers. However, abuse of customers' trust can mean a loss of gained competitive advantage.

E-commerce designates a huge freedom when shopping and offers access to cheaper products from all over the world (thanks to economies on the transaction costs). Theoretically, trust in enterprise-customer relations should be laid out equally. On the Internet, there exists an unlimited number of buyers and vendors, which indicates a resemblance to a free competition market. The number of business transactions should be therefore mountainous. Practice however denies this. There comes to light the fact that a major limitation on the development of Internet trade comes from mistrust. The skeptical approach of customers to products offered via the Internet confirms netizens' opinions (Culnan and Armstrong 1999). Trust building in electronic commerce is an assignment considerably more difficult than in the traditional offline trade. The barrier lies in the lack of personal contact between transaction parties. A businessman working through the Internet is not perceived as stable and firm. In an online market, there are not any barriers to entry and exit, so tradesmen can very quickly appear and even more quickly disappear. In traditional trade there exist certain obstacles connected with immovables, such as staff. In e-commerce, there is anxiety for the loss of privacy, which is connected with the usage of the credit card what can cause unwanted utilizations of the information about consumer necessary to complete the transaction. The lack of personal contact and the loss of privacy are two main sources of concern pointed out by citizens of the country with the most advanced "new economy," the United States of America (Hoffman et al. 1999).

According to Kossecki (n.d.), the main objectives of trust building in e-commerce are:

- Reduction of subjective and objective uncertainty and the risk of the contracting party connected with the transaction
- Conviction of the potential customer to do the shopping
- Enlargement of the loyalty of already-gained customers
- Enhancement of participation of Internet shop in expenses of the customer, both by the diminution of the quantity of shopping made by means of traditional distribution channels and also done in other online shops
- Enhancement of the relation with customers
- Improvement of customer satisfaction
- Data collection in order to strengthen relations with customers.

The above targets can be divided into two groups. In the first group of aims, the priority is to persuade the customer to make a purchase via the Internet (online). The second group of purposes refers to intensifying this relation with the customer and winning loyalty.

An example of an initiative supporting trust building in B2C relations is Euro-Label, the European sign of trust for consumers and salesmen in e-commerce (http://www.euro-label.com/). The sign "Euro-label" on a website means that the buyer will receive trustworthy and reliable service and that:

- The salesman presenting the offer online exists in reality
- The terms of sale are intelligible and described on the shop window
- The salesman takes into consideration regulations about the protection of personal data
- The product will be delivered as ordered
- In a situation of irregularity in the transaction, there will be a procedure for making a complaint.

Table 2.3 lists institutions marked with the Euro-label in different countries from the European Union.

In order to receive Euro-Label, a vendor must adopt the European Code of Practice and meet the legal requirements. The institution issuing Euro-Label (certification body) is responsible for checking that the salesman meets the above criteria. Copyright and suitable mechanisms of confirmations assure that the sign, visible on the website vendor, is reliable.

To draw nearer the trust issues in enterprise-customer relations, we present below the chosen, most important models explaining the occurrence of B2C trusting

TABLE 2.3
Certification Bodies for Euro-Label

Country	Certification Bodies for Euro-Label
Austria	Das Österreichische E-Commerce-Gütezeichen http://www.guetezeichen.at/
France	Label site http://www.labelsite.org/
Spain	Confederacion espanola de comercio/Confianza Online http://www.confianzaonline.org/
Germany	Geprüfter Online-Shop http://www.shopinfo.net/
Poland	Instytut Logistyki i Magazynowania http://www.ilim.poznan.pl/
Italy	ConfCommercio http://www.euro-label.org/

relations. The acquaintance of the model admits us to use it in order to build trust. That is why trust building in relations between enterprise and customer follows modelings of this occurrence. Often, research is carried out with fictitious, specially designed websites, which can decrease the reliability of a model built in this way (e.g., Stewart 2003). The number of models constructed on the grounds of findings is limited (Lee and Turban 2001; McKnight, Choudhury, and Kamar 2002).

Trust Model in Relationships between Subjects in Cyberspace

The process of building and developing trust in relationships between company and consumer and between two companies is also recognized as a process of trust production (Bailey, Gurak, and Konstan 2002). Presented here, the trust model in relationships between subjects in cyberspace refers also to the relationship described in detail in the section "Trust Building in Relations between Business and Business (B2B)."

The trust-production process can be summarized in three main steps:

Step 1—The trustee produces extrinsic cues to convey trustworthiness.
Step 2—The trustor perceives the extrinsic cues produced by the trustee and elsewhere.
Step 3—Based on perceived cues and internal beliefs, the trustor attributes a level of trustworthiness to the trustee. (Bailey, Gurak, and Konstan 2002)

The attribution of trust appearing in Step 3 may be greater than, less than, or equal to the trustee's intrinsic value of trustworthiness, each case having an essential implication for the economic exchange (Brainov and Sandholm 1999).

In the trust-production process, extrinsic trust cues are produced by various multiple sources and perceived along with several dimensions that lead to trust attribution. The trustor confides in the trustee using different types of signs, trust sources. Each **trust source** stands for a belief, impression, experience, or institution by which a trust cue is compelled (Bailey, Gurak, and Konstan 2002). According to the literature, four main sources of trust are derived. These studies cover literature in human-computer interaction (Fogg and Tseng 1999, Kim and Moon 1997, Tseng and Fogg 1999), social psychology (Deutsch 1973, Giffin 1967, Patzer 1983, Rempel, Holmes,

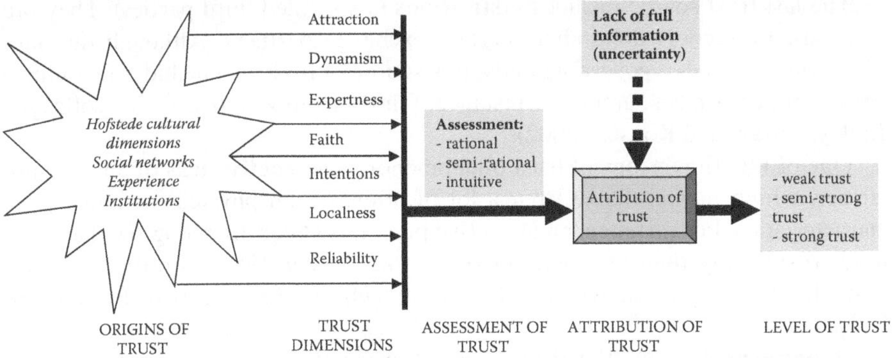

FIGURE 2.11 The trust model of relationships between subjects in cyberspace. Source: Partly adapted from Bailey, B. P., Gurak L. J., and Konstan, J. A. 2002. Trust in cyberspace. In Human factors and web development, 2nd ed., J. Ratner (ed.), 311–321. Mahwah, NJ: Lawrence Erlbaum Associates.

and Zanna 1985, Rotter 1980), marketing (Doney and Cannon 1997, Johnson and Grayson 2000), and economics (Zucker 1986). The occurrence of all or only some trust sources is not constant, but rather it is changeable in time.

At the same time, a **trust dimension** is a reflection of a feature of trust that is pivotal in the decision to trust or not trust somebody. A person cognitively combines the trust dimensions forming an overall attribution of trust. Trust sources and trust dimensions are tightly connected and related to each other (i.e., a single trust source may contribute to multiple dimensions of trust, whereas a single dimension of trust may be contributed to shaping multiple sources) (Bailey, Gurak, and Konstan 2002). In the trust model of relationships between subjects in cyberspace, there are separated seven dimensions of trust based on the analysis of social psychology literature (Giffin 1967; Patzer 1983; Posner and Kouzes 1988; Rempel, Holmes, and Zanna 1985).

The first source of trust is **presumptions.** The process of trust warranting strongly originates from general assumptions and stereotypes existing within a certain culture or level of trust maintained when there are no doubts. For example, a partner can perceive another partner as more or less trustworthy because of its location or style and colors used in its offices (photos can be available to be seen online) and on its Web site.

Then we have **surface inspection,** which is responsible for trust production through an examination of an exchange partner's external appearance, such as the visual design of a Web site or the physical appearance of workers. We need to remember that, once formed, first impressions are extremely difficult to dismiss or change (Zajonc 1980).

A further source is **experience,** which refers to previous successful exchanges with a transaction partner. For example, earlier repeated successful transactions with an exchange partner cause the image of this partner to be more trustworthy than an unfamiliar partner. Reputations, opinions from other partners familiar to us, and brands can also be taken into careful consideration as an experience-based source of trust (Zucker 1986).

The last trust source includes **institutions** (also called third parties). They produce trust by records done about certain exchange partners. Although the name "institution" implies a governing body, this source of trust also includes recommendations from various sources, for instance, family members, friends, or colleagues (Bailey, Gurak, and Konstan 2002).

One of the dimensions of trust phenomenon is **attractiveness** of an exchange partner, which means that we put a lot of attention on physical or nonphysical characteristics. For instance, an attractive person is generally recognized as being more trustworthy than an unattractive person (Patzer 1983). Similarly, an aesthetically pleasing exchange site should be seen as more trustworthy than one that is not.

Dynamism of the additional communication channels is provided by an exchange partner through various ways, such as oral, written, or visual. An example of dynamism can be information about a partner's Web site on a ticker-tape display.

By **expertise,** we understand a partner's needful and relevant skill, ability, or specific knowledge, which cannot be overestimated in cyberspace. An expert is usually recognized as being more trustworthy than a nonexpert, so expertness affects trust (Brainov and Sandholm 1999; Peters, Covello, and McCallum 1997).

Obligations' fulfillment by an exchange partner in unknown future conditions is included in **faith.** Faith plays a very special role in situations when there are no previous experiences or when those experiences are only minimal and not sufficient. For instance, when buying a computer via an Internet vendor, we hope that the computer will work well and we have faith that this purchase will be a good decision.

Detectable and foreseen targets and objectives of an exchange partner are included in dimension called **intentions.** For example, a seller who shows openness and honesty and aims to secure the welfare of the buyer is recognized as being more plausible (Doney and Cannon 1997).

Another significant dimension of trust is the **proximity** of an exchange partner's ideas, beliefs, values, or geography. The exchange partner who is located closer to our place or who is involved in charity is worthier of trust.

The seventh trust dimension is the **reliability** of an exchange partner, which can be estimated and measured in terms of dependability, predictability, and consistency. Reliability is strictly established on past experiences and prior interactions. For instance, a shopper who purchased a book online and was satisfied with the transaction will see that seller as more reliable than an unknown seller.

Customers develop trust in an exchange partner from cyberspace by taking into account all described sources and dimensions of trust. The trust model presented above can be employed in many types of relationships between partners, not only in B2C.

Trust Model for Social Relations between Multichannel Retailer and Customer

The trust model for social relations between multichannel retailer and customer built by Kuan and Bock (2005) draws attention to the importance of relations and various social networks. A multichannel retailer stands for a vendor acting simultaneously on traditional market and online market (e-commerce). The authors of this trust model explained all elements step by step by hypothesizing. Very special emphasis is put on trust based on affection.

Hypothesis 1: The level of a customer's emotional trust is positively related to the level of the customer's perceived continuity of online relationship (Kuan and Bock 2005).

Hypothesis 2: The level of the customer's cognitive trust of the online vendor is positively associated with the level of emotional trust in the online vendor (Kuan and Bock 2005).

Different forms of social capital, such as information channels, norms and sanctions, and a company's obligations toward customers, influence cognitive trust level (Coleman 1988). That is why very essential here are positive opinions within existing social relations and networks, which are the fundamental source of information for customers (Granovetter 1973). This is possible using for instance various kinds of chat forums, which are recognized as the platforms for exchange of information about specific products.

Hypothesis 3a: The availability of positive referrals in offline social networks is positively related to the level of a customer's cognitive trust of the online vendor (Kuan and Bock 2005).

Hypothesis 3b: The availability of positive referrals in online social networks is positively related to the level of a customer's cognitive trust of the online vendor (Kuan and Bock 2005).

Effective norms and sanctions should keep down all kinds of frauds in e-commerce and, in consequence, this decreases the fear of dishonest actions of Internet vendors. It can lead to punishing online vendors voided of integrity. Customers can share with each other negative opinions what can cause finally the drop in vendor turnovers (Jones et al. 1997).

Hypothesis 4a: The level of online purchasing norm is positively related to the level of a customer's cognitive trust in an online vendor (Kuan and Bock 2005).

Hypothesis 4b: The availability of sanctions is positively related to the level of a customer's cognitive trust in an online vendor (Kuan and Bock 2005).

A company is obliged to meet all customer expectations. The level of trust in a vendor is affected by the customers' feelings in this aspect. Quality of service is determined by comparison between what is expected by the customer and what is provided.

Hypothesis 5a: The level of perceived offline service quality is positively related to the level of a customer's cognitive trust in the online vendor (Kuan and Bock 2005).

Hypothesis 5b: The level of perceived online service quality is positively related to the level of a customer's cognitive trust in the online vendor (Kuan and Bock 2005).

According to the Social Network Theory, when the strength of the ties between a customer and a source of positive information is strong, the relationship between the access to positive referrals and the level of cognitive trust in customers' opinions can be more positive (Granovetter 1973).

Hypothesis 6a: The perceived strength of ties with referral sources in the customer's offline social network is a moderator of the positive relationship between the availability of positive offline referrals and the level of the customer's cognitive trust in the online vendor (Kuan and Bock 2005).

Hypothesis 6b: The perceived strength of ties with referral sources in the customer's online social network is a moderator of the positive relationship between the availability of positive online referrals and the level of the customer's cognitive trust in the online vendor (Kuan and Bock 2005).

When the level of source centrality is higher, the relationship between the availability of positive referrals and the cognitive trust of the customer in the online vendor is more apparent. All information received about the online vendor's behaviors are recognized as more unprecarious and reliable.

Hypothesis 6c: The level of perceived centrality of referral sources in the customer's offline social network is a moderator of the positive relationship between the availability of offline positive referrals and the level of the customer's cognitive trust in the online vendor (Kuan and Bock 2005).

Hypothesis 6d: The level of perceived centrality of referral sources in the customer's online social network is a moderator of the positive relationship between the availability of online positive referrals and the level of the customer's cognitive trust in the online vendor (Kuan and Bock 2005).

Offline and online service quality is based on the customers' direct, social relations with the retailer as well as the purchasing experience. Both trust literature and empirical studies have proven that customers' purchasing experiences influence their emotions (e.g., Taylor 2000).

Hypothesis 7a: The perceived quality of offline service is positively related to the level of the customer's emotional trust in the online vendor (Kuan and Bock 2005).

Hypothesis 7b: The perceived quality of online service is positively related to the level of the customer's emotional trust in the online vendor (Kuan and Bock 2005).

Taking into account not only sociological theories but also practical results of research, Kuan and Bock created the trust model to be used in research. This research tool lets us explore the phenomenon of trust in the context of social relations as well as networks of customers.

All of the above hypotheses are included in questionnaire form in order to put theory into practice. The Trust Model for Social Relations between Multichannel Retailer and Customer by Kuan and Bock is a great step toward better understanding the aspect of affective trust. Trust in this model is explored in the context of multichannel retailers and customers.

An Integrative Model of Consumer Trust in Internet Shopping

The starting point for Cheung and Lee (2003) to create an integrative model of consumer trust in Internet shopping is modification of the definition of trust proposed by Mayer et al. (1995). For this model is used the following definition of consumer trust in Internet shopping (Lee and Turban 2001): "the willingness of a consumer to be vulnerable to the actions of an Internet merchant in an Internet transaction, based on the expectation that the Internet merchant will behave in certain agreeable ways, irrespective of the ability of the consumers to monitor or control that Internet merchant" (p. 79).

Shopping via the Internet is more and more common nowadays; it replaces traditional offline trade. There are many very apparent differences between these two forms of trade. First of all, the buyer cannot physically check the quality of the acquired product, as is accomplishable in offline shopping. The Integrative Model of Consumer Trust in Internet Shopping according to Cheung and Lee (2003) intends to explain the phenomenon of trust in these circumstances.

The Integrative Model of Consumer Trust in Internet Shopping according to Cheung and Lee (2003) recounts perceived trustworthiness of an Internet vendor, the external environment, and perceived risk as the three major determinants of consumer trust in Internet shopping. Moreover, propensity to trust exhibits a moderating impact on the relationships between consumer trust and its antecedents (except perceived risk).

Adapting the social psychology perspective of trust (Lewicki and Bunker 1995), Cheung and Lee believed that **trustworthiness** is one of the key determinants of consumer trust in Internet shopping. In their study, they identified four trustworthiness factors that are specific and unique to all Internet vendors: perceived integrity (PI), perceived competence (PC), perceived security control (PSC), and perceived privacy control (PPC).

PI embraces online shoppers' perceptions of Internet vendors' honesty. We consider here if the vendor performs consistent actions, whether its actions are coherent with what was promised or said, and whether its transactions with its consumers are done fairly. **PC** represents Internet shoppers' perceptions of Internet vendors' skills and expertise. **PSC** is understood as Internet shoppers' perceptions of Internet vendors' abilities to fulfill security requirements stated in the transaction agreement. **PPC** is recognized as Internet shoppers' perceptions of Internet vendors' abilities to protect from unauthorized use or disclosure of consumers' personal information collected from its online transactions (Cheung and Lee 2003).

When considering this trust model, we need to take into account a very important element: the **external environment.** In sociology, trust is seen as the ability to structure contracts or rewards and punishments and in consequence to make units behave in a prespecified, more predictable manner. This kind of trust is a significant source for maintaining social orders (Lewis and Weigert 1985). That is why in the virtual world it is a very essential issue.

Another element in this trust model is **third-party recognition** (TPR). TPR explains Internet shoppers' perceptions of the effectiveness of TPR bodies in assuring the trustworthiness of Internet vendors. Moreover, Internet vendors are acting within a certain law environment. This brings us to the next factor, the **legal framework** (LF), which is the perceived effectiveness of the law and code of practice established to protect Internet shoppers during electronic transactions.

We need to emphasize that the phenomenon of trust has its origins in the occurrence of risky situations. In the beginning, we face risk and then it is followed by trust (Boon and Holmes 1991; Mitchell 1999). **Perceived risk** (PR) is conceptualized as Internet shoppers' perceptions of the possibility of yielding unexpected outcomes or even negative consequences (Cheung and Lee 2003).

Propensity to trust (PT) is the last element in Cheung and Lee's integrative model. It is recognized as a personality trait shaping generalized expectations of trustworthiness of other people, objects, situations, and so forth. PT is defined by cultural background, personality type, and developmental experiences (Hofstede 1980). Considering consumer trust in Internet shopping, one must admit that a consumer's propensity to trust has a big influence on the relationship between trust and its antecedents.

All of the elements of this model appear in the following formula (Cheung and Lee 2003):

$$\begin{aligned} &\text{Trust in relations B2C} = \\ &= PSC + PPC + PI + PC + PT + TPR + LF + PR + \\ &+ (PSC*PT) + (PPC*PT) + (PI*PT) + \\ &+ (PC*PT) + (TPR*PT) + (LF*PT) \end{aligned} \tag{2.1.1}$$

The Integrative Model of Consumer Trust in Internet Shopping represents a holistic approach and an explanation of the trust issue in cases of online shopping. We cannot forget that the phenomenon of trust is interdisciplinary and is being studied and explored not only by economists but also by sociologists, psychologists, and so on. The main point in the model is perceived risk. Perceived risk of Internet shopping is negatively associated with consumer trust in Internet shopping. The higher the risk, the lower the propensity to complete the transaction in B2C relationships. This risk determines consumer trust. All information in this model is very valuable from vendors' points of view and should be used as guidelines for future Internet shopping transactions.

Furthermore, primordial for consumer trust in Internet shopping is clarity. It is to be built by the good image of an Internet vendor. Fiore (2001) reckoned that all information about online transactions, the pricing policy, and so on must be clear and well understood by shoppers. "Consumers don't like surprises. If you promote your site as the low-price leader, your prices should show it. If you're a boutique shop and charge higher-than-average prices, show the value you've added to your products or service. Make it very clear what your charge and why – and make sure it fits the expectations of your site's visitors" (Fiore 2001).

The final conclusion from the integrative model is that only companies that can effectively manage their consumers' online shopping experiences and build their consumer loyalty will eventually survive and will be trusted by shoppers.

REFERENCES

Bailey, B. P., Gurak L. J., and Konstan, J. A. 2002. Trust in cyberspace. In *Human factors and web development*, 2nd ed., J. Ratner (ed.), 311–321. Mahwah, NJ: Lawrence Erlbaum Associates.

Boon, S. D., and Holmes, J. G. 1991. The dynamics of interpersonal trust: Resolving uncertainty in the face of risk, *Cooperation and Prosocial Behavior*, R. A. Hinde and J. Groebel (eds.), 190–211 Cambridge: Cambridge University Press.

Brainov, S., and Sandholm, T. 1999. Contracting with uncertain level of trust. Proceedings of ACM Conference on Electronic Commerce.

Caccioppo, J., and Petty, R. 1981. *Attitudes and persuasion: Classic and contemporary approaches*. New York: Brown Company Publishers.

Cheung, Ch. M. K., and Lee, M. K. O. 2003. An integrative model of consumer trust in Internet shopping. In Proceedings of the 11th European Conference on Information Systems, 479–492. ECIS 2003, Naples, Italy, 16–21 June 2003.

Coleman, J. S. 1988. Social capital in the creation of human capital. *American Journal of Sociology*, 94(Suppl.), 95–120.

Culnan, M. J., and Armstrong, P. K. 1999. Privacy concerns, Procedural fairness, and impersonal trust: An empirical investigation. *Organization Science*, 10, 104–115.

Doney, P. M., and Cannon, J. P. 1997. An examination of the nature of trust in buyer-seller relationships. Journal of Marketing, 61(2), 35–51.

Euro-Label. n.d. http://www.euro-label.com/ (accessed August 7, 2007).

Fiore, F. 2001. The "Big Five" of online shopping, *informIT*, Nov 09.

Fogg, B. J., and Tseng, S. 1999. The elements of computer credibility, 80–87. Conference on Human Factors in Computing Systems. Proceedings of the SIGCHI conference on Human factors in computing systems: The CHI is the limit. New York: ACM Press.

Fukuyama, F. 1995. *Trust: The social virtues and the creation of prosperity*. New York: Free Press Paperbacks.

Giffin, K. 1967. The contribution of studies of source credibility to a theory of interpersonal trust in the communication process. *Psychological Bulletin*, 68(2), 104–120.

Granovetter, M. 1973. The strength of weak ties. *American Journal of Sociology*, 78(6), 1360–1380.

Hoffman, D. L., Novak, T. P., and Peralta, M. A. 1999. Building consumer trust online. *Communication of the ACM*, 42(4), 80–85.

Hofstede, G. 1980. Motivation, leadership, and organization: Do American theories apply abroad? *Organizational Dynamics*, 9(1), 42–63.

Hoyer, W. D., and Brown, S. P. 1990. Effects of brand awareness on choice for a common, repeat-purchase product. *Journal of Consumer Research*, 17, 822–829.

Johnson, D. S., and Grayson, K. 2000. Sources and dimensions of trust in service relationships. In *Handbook of services marketing and management*, T. A. Swartz and D. Iacobucci (eds.), 357–370. Thousand Oaks, CA: Sage Publications, Inc.

Jones, C., Hesterly, W. S., and Borgatti, S. P. 1997. A general theory of network governance: Exchange conditions and social mechanisms. *Academy of Management Review*, 22(4), 911–945.

Kim, J., and Moon, J. Y. 1997. Designing towards emotional usability in customer interfaces: Trustworthiness of cyber-banking system interfaces. *Interacting with Computers*, 10, 1–29.

Kossecki, P. n.d. Budowa zaufania klientów w handlu elektronicznym. http://www.e-marketing.pl/artyk/artyk83.php (accessed August 7, 2007).

Kuan, H. H., and Bock, G. W. 2005. The collective reality of trust: An investigation of social relations and networks on trust in multi-channel retailers. In Proceedings of the 13th European Conference on Information Systems, Information Systems in a Rapidly Changing Economy. ECIS 2005, Regensburg, Germany, May 26–28, 2005.

Lee, M. K. O., and Turban, E. 2001. A trust model for consumer Internet shopping. I*nternational Journal of Electronic Commerce*, 6(1), 75–91.

Lewicka-Strzałecka, A. 2003. Zaufanie w relacji konsument – biznes. Prakseologia, 12, 195–207.

Lewicki, R. J., and Bunker, B. B. 1995. Trust in relationships: A model of trust development and decline. In *Conflict, cooperation, and justice*, B. Bunker and J. Rubin (eds.), 133–173. San-Francisco: Jossey-Bass.

Lewis, J. D., and Weigert, A. 1985. Trust as a social reality. *Social Forces*, 63(4), 967–985.

McKnight, D. H., Choudhury, V., and Kamar, C. 2002. Developing and validating trust measures for e-commerce: An integrative typology. *Information Systems Research*, 13(3), 334–359.

Mitchell, V. W. 1999. Consumer-perceived risk: Conceptualizations and models. *European Journal of Marketing*, 33(1/2), 163–195.

Moorman, C., Zaltman, G., and Deshpande, R. 1992. Relationships between providers and users of market research: The dynamics of trust within and between organizations, *Journal of Marketing Research*, 29, 314–328.

Patzer, G. L. 1983. Source credibility as a function of communicator physical attractiveness. *Journal of Business Research*, 11(2), 229–241.

Peters, R. G., Covello, V. T., and McCallum, D. B. 1997. The determinants of trust and credibility in environmental risk communication: An empirical study. *Risk Analysis*, 17(1), 43–54.

Posner, B. Z., and Kouzes, J. M. 1988. Relating leadership and credibility, *Psychological Reports*, 63(2), 527–530.

Rempel, J. K., Holmes, J. G., and Zanna, M. P. 1985. Trust in close relationships, *Journal of Personality and Social Psychology*, 49(1), 95–112.

Rotter, J. B. 1980. Interpersonal trust, trustworthiness, and gullibility. *American Psychologist*, 35(1), 1–7.

Stewart, K. J. 2003. Trust transfer on the World Wide Web. *Organization Science*, 14(1), 5–17.

Taylor, R. K. 2000. Marketing strategies: Gaining a competitive advantage through the use of emotion. *Competitiveness Review*, 10(2), 146–152.

Uzzi, B. 1999. Embeddedness in the making of financial capital: How social relations and networks benefit firms seeking financing. *American Sociological Review*, 64(4), 481–505.

Zajonc, R. B. 1980. Feeling and thinking. *American Psychologist*, 35(2), 151–175.

Zucker, L. G. 1986. Production of trust: Institutional sources of economic structure. 1840–1920. *Research in Organizational Behavior*, 8, 53–111.

2.5 TRUST BUILDING IN RELATIONS BETWEEN BUSINESS AND BUSINESS (B2B)

Trust building in B2C relationships differs from trust building in the second model of relationships, between business and business (B2B), not only due to the parties involved in this process but mainly due to the degree of complexity in these relationships. Consumer trust refers to some client or a group of clients and a certain enterprise. On the contrary, trust in B2B relationships refers to various types of companies and their employees. Processes (for instance, selling semifinished products and raw materials) between economic subjects are definite parts (elements) of a value chain and especially of a supply chain (Handfeld and Bechtel 2002). In each economic subject there are many people involved in these types of processes. Furthermore, it is required in B2B relationships to have an agreement to do some certain activities or to make some certain decisions. A good example of this is a fixed level of orders within which employees do not need to ask their superiors for permission.

Another difference between B2C and B2B models in e-commerce is the volume of turnovers (Graph 2.1) presented according to the estimates and forecasts from Forrester, IDC, eMarketer, and Gardner Group.

According to diverse sources, B2B increased from between $278 and $433 billion in 2000 to between $4,320 and $8,823 trillion in 2005. This growth is said to be still growing very dynamically and it is believed to gain the record-breaking level of $12 trillion in 2006.

The B2B sector is much more extensive and large (Laudon and Traver 2002) than the B2C sector (Graph 2.2). Graphs 2.2, 2.3, and 2.4 present the B2B and B2C scenarios in 2006 for different regions: North America, Asia/Pacific, West Europe, Latin America, East Europe, and Africa/Middle East in billions of dollars as well as in percentages. Both B2B and B2C global e-commerce are expected to remain very high in North America (respectively, 58.1% and 37.6%). The other regions are going to have smaller shares in B2B and B2C markets. Behind North America are Asia/Pacific (respectively for B2B it is 20% and for B2C 33%) and West Europe (respectively 18.9% and 24.7%). The last three positions, with shares smaller than 3%, are held by Latin America, East Europe, and Africa/Middle East. What is so surprising is that, in the very early phases, B2C was an engine that redirected the commerce to the Internet and then B2C was replaced by B2B. We can notice that fact analyzing the proportion between B2B and B2C. Moreover, the average value of a purchase via the Internet in B2B fluctuates between $50,000 and $250,000, whereas in B2C it is $75 (Phillips and Meeker 2000, p. 23).

Trust is sometimes named the "Achilles heel" with reference to accepting and choosing an e-market (electronic marketplace) (Hsiao 2003). A **public electronic market** stands for:

- Electronic space open to several buyers and several sellers
- Trading platform
- A market which itself does not sell or buy goods or services traded on the platform but has a trading function to be a moderator between interested and involved parties

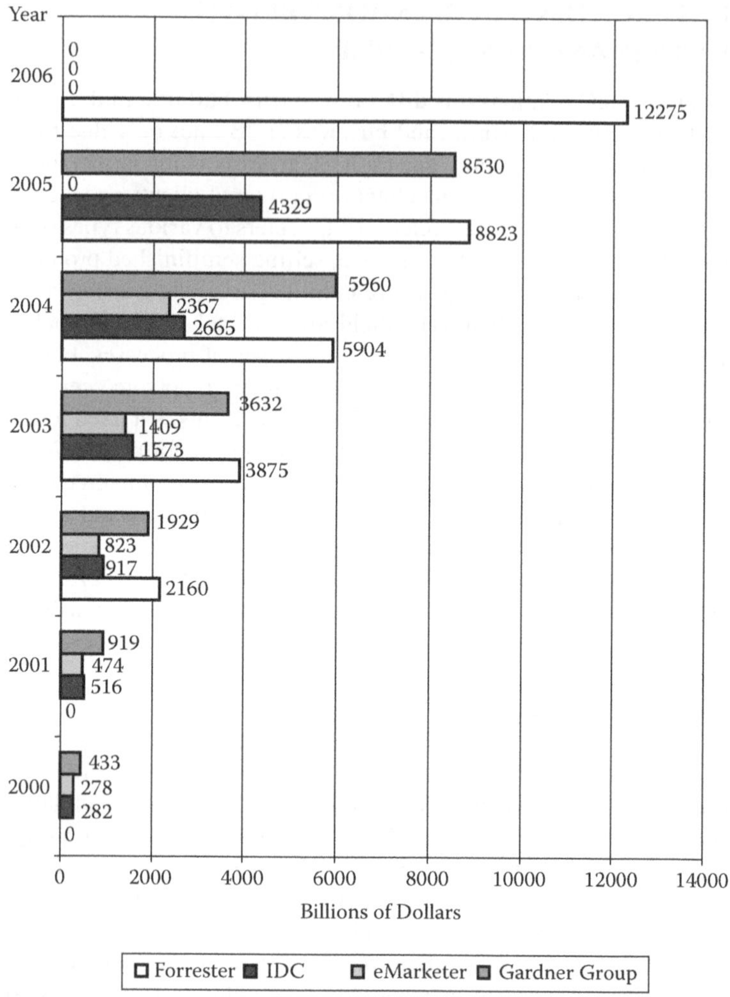

GRAPH 2.1 Worldwide B2B: Some global estimates and forecasts (billions of dollars). Source: Adapted from E-commerce and development report 2002. New York and Geneva: United Nations, p. 7, and Singh, V. 2004. E-marketplace development. Report from eMarketServices. http://www.emarketservices.com (accessed August, 8, 2007), p. 2.

- A market where many buyers and sellers with and without previous experience of mutual collaboration come to initiate and complete a transaction
- A market often called the "B2B Internet platform." (eMarketServices 2004b; Grieger 2003)

E-market is precisely the point where we need to mention supplier directories. They are focused on B2B and B2C. The aim of supplier directories is to support companies in establishing new business relationships. However, there is no actual trade on these platforms. It is not a B2B e-market or supplier directory if it:

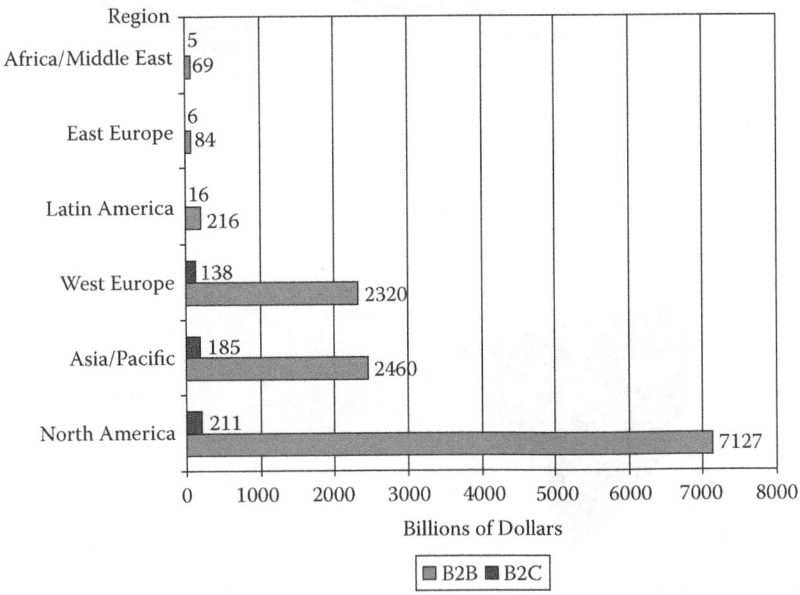

GRAPH 2.2 B2B and B2C forecast in 2006 by region (billions of dollars). Source: Adapted from E-commerce and development report 2002. New York and Geneva: United Nations, p. 8, and Singh, V. 2004. E-marketplace development. Report from eMarketServices. http://www.emarketservices.com (accessed August, 8, 2007), p. 3.

- is a Web site of one company selling its own products
- is one seller acting as a distributor, selling other companies' products
- mainly focuses on trade to or between consumers
- is only a provider of e-commerce solutions
- focuses on nontrading activities like collaborative design or people collaboration. (http://www.emarketservices.com/, eMarket Basics)

The number of public electronic markets grows each year. We can distinguish many electronic marketplaces from these covering very wide, extensive branches to those very narrowly specialized ones. They operate in many different branches and sectors, for example:

- Achilles—energy and fuels, transportation, construction
- Atento—office equipment, mobile phones, transportation, furniture
- Endorsia and MundoAcero—metal and mining
- FORDAQ—forestry and wood
- Global Healthcare Exchange—health care and pharmaceuticals
- Global Wine and Spirits, Pro-Wine, and Vertical Wine—wines
- Impexbobiz—multiple industries, from agriculture to cosmetics to software
- Logitall—IT products, telecommunication and office equipment
- SupplyOn—automotive products

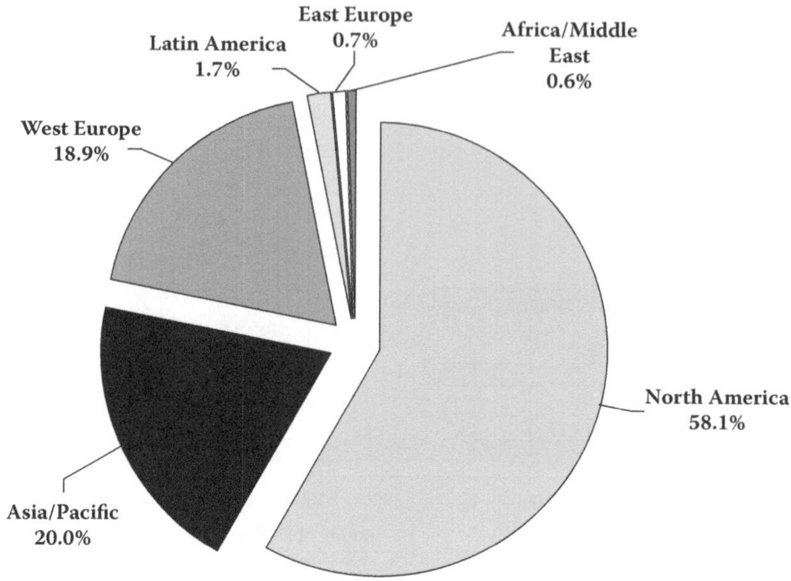

GRAPH 2.3 B2B forecast in 2006 by region (%). Source: Adapted from E-commerce and development report 2002. 2002. New York and Geneva: United Nations, p. 8, and Singh, V. 2004. E-marketplace development. Report from eMarketServices. http://www.emarketservices.com (accessed August, 8, 2007), p. 3.

- The Nordic Power Exchange—electricity and more
- Tradeplace—electronics and electrical products. (Kjølseth 2006; McVey 2005, Zällh 2005)

These above-selected examples show the growing importance of electronic marketplaces all over the world and in many various branches and sectors.

Factors determining the decision about a purchase, its size, and repeatability are associated with the buyer. The buyer is recognized as an economic subject (transactor) or an individual. In B2C relationships, a crucial role is played by the level of price, which should be more competitive comparing to others. At the same time, the success of trust building in B2B relationships is affected by, for instance, quality of a product or service, reliability, fixed deliveries on time, and accepted and well-known relationships between the exchange partners (Raisch 2001). Enterprises might also specify more particular requirements that need to be met by their business partners in order to develop trusting relationships and continue collaboration in the future. These additional conditions are associated with the characteristics of each sector or branch—for instance, it can be not only high quality of products but also the fact that they are environmentally friendly.

The partner selection process heavily influences the trust-building process. The reasons for selecting a specific partner for collaboration can be as follows: the value of fixed-assets balance, a perspective for creating new value and friendly relationships during cooperation. In B2B relationships, the factors recognized as most

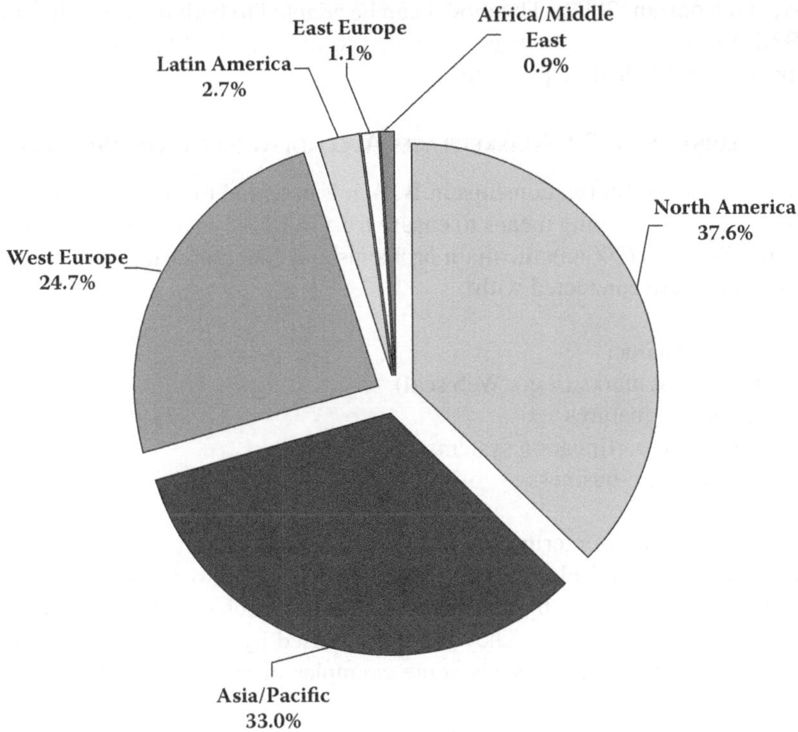

GRAPH 2.4 B2C forecast in 2006 by region (%). Source: Adapted from E-commerce and development report 2002. New York and Geneva: United Nations, p. 8, and Singh, V. 2004. E-marketplace development. Report from eMarketServices. http://www.emarketservices.com (accessed August, 8, 2007), p. 3.

important are brand, reputation, advertising, and price (Phillips and Meeker 2000, p. 23). The complexity of B2B relationships is described by various factors. To be more specific, Schneider (2002, p. 181) found that processes between companies consist of 20 different steps. Then, he grouped them together and received three major groups, which are as follows:

1. Searching for potential business partners
2. Partners' assessment, setting the price level, and the source of required supplies
3. Completing the purchase transaction, settling accounts, and payment. (EU Commission, 2002)

These three groups of activities in processes occurring between partners in B2B relations reflect only some actions from the whole set of actions. They concentrate on adjusting the delivery to buyers' requirements.

In the section "Trust Building in Relations between Business and Client (B2C)" is presented a trust relationship model for subjects in cyberspace according to Bailey,

Gurak, and Konstan (2002). The model can be adapted to both relationships, in B2C and B2B. In this section, other ideas related to the trust-building process between enterprises in B2B shall be presented.

Trust Building in B2B E-Marketplaces According to eMarketServices

For creating successful relationships in B2B, it is essential to explore the nature and extent of available, existing means to ensure trust in B2B e-marketplaces. According to eMarketServices (2004a), the main problems and concerns affecting trust in B2B e-marketplaces are connected with:

- codes of conduct
- various trust marks (logo, Web seal)
- electronic signatures
- security and certification systems
- standards for e-business.

Code of conduct is a term used for a service standard that we expect to receive in certain transactions dealing with an organization. Moreover, it defines what conditions need to be fulfilled by the Internet vendor. Codes of conduct are only recognized as sets of guidelines. They are not imposed by law and are not legislated (Kjølseth 2005). Table 2.4 presents some examples of codes of conduct in e-commerce and e-business.

Trust marks (logo, Web seal) take a form of a symbolic representation, for instance, a logo issued by an organization auditing the online operation in a certain region. They are placed at electronic marketplaces or Web sites of certain online vendors to demonstrate specific standards or criteria fulfillment. As a consequence, potential partners are more likely to put their trust in the online vendor.

According to the definition given by the U.S. Electronic Signatures in Global and National Commerce Act of 2000 (ESIGN Act, Section 106. Definitions, 114 STAT 472), an **electronic signature** is defined as: "an electronic sound, symbol, or process attached to or logically associated with a contract or other record and executed or adopted by a person with the intent to sign the record." An electronic signature simplifies some processes and transactions and saves time. The main benefits of an electronic (digital) signature are:

- it is equal to a traditional, handmade signature (written form)
- it is only connected with and made by the party making a signature (signer)
- it allows identification of the signer
- according to the legal significance of the signer's act, there is no need for the signer's physical presence during most legal transactions
- it requires no need for storage. (Atena 2005)

Security and certification systems cover all activities and monitoring associated directly with continuously satisfying the conditions described in codes of conduct. It is adapted in all types of economic activity, regardless of its ownership form (Vargas

TABLE 2.4
Examples of Codes of Conduct in E-Commerce and E-Business

E-Commerce	
Alufoil – European Aluminium Foil Association	http://www.alufoil.org/media/eCommerce.pdf
ArGeZ (Arbeitsgemeinschaft Zulieferindustrie) Fairness for Electronic Market Places	http://www.argez.de/
CPGmarket.com (Nestle, Danone, Henkel, SAP) – Consumer Packaged Goods Industry	http://www.cpgmarket.com/index.htm
Europacable (European Confederation of Associations of Manufacturers of Insulated Wires and Cables)	http://www.europacable.com/
E-Business	
Electronic Commerce Promotion Council of Japan (ECOM) – ECOM Guidelines for Transactions between Virtual Merchants and Consumers	http://www.ecom.or.jp/ecom_e/report/no7/wg01.html
FEDMA code on e-commerce and interactive marketing	http://www.fedma.org/code/page.cfm?id_page=106 http://www.fedma.org/img/db/ Code_of_conduct_for_e-commerce.pdf
ECP (Electronic Commerce Platform, Netherlands) Model Code of Conduct for Electronic Business	http://www.ecp.nl/index.php
Canadian Association of Internet Providers (CAIP) – "Shopping Safely Online" guide has a brief mention of online auctions	http://www.caip.ca/issueset.htm
New Zealand Direct Marketing Standards Authority—Code of practice for direct marketing in New Zealand	http://www.trustmark.co.nz/Story?Action=Viewandid=1297

Source: Kjølseth, B. B. 2005. Applying codes of conduct and trust marks. Report from eMarketServices. http://www.emarketservices.com (accessed August, 8, 2007). With permission.

2005). Application of security and certification systems is not required or imposed, but it has a significant impact on building trust relationships in B2B and B2C.

Chambertrust and Chambersign present two examples of very important initiatives in the domain of security and certification systems. Chambertrust is an initiative set up by the Ministries of Trade or Economy and Chamber of Commerce in compliance with generally accepted guidelines in order to build trust in business relationships. Chambertrust is run by the World Chambers Federation (WCF), which is a highly specialized department in the International Chamber of Commerce (ICC), associating members from every country (one member from each country). Chambertrust verifies whether a certain organization exists, what its activities range is, and what products or services it sells in comparison with its description in its online trade offer. Chambertrust helps independent small- and mid-sized enterprises to overcome trust obstacles, which an enterprise faces outside its local market beyond which it is essentially "brandless" and

TABLE 2.5
Examples of Codes of Conduct

Names of Codes of Conduct	Source	Conduct
BBB Reliability Seal and BBB Privacy Seal	http://www.bbbonline.org/ U.S.	Securing suitable level of quality, privacy, and content coherent with ethical principles
TRUSTe	www.truste.org Founder Electronic Frontier Foundation and CommerceNet	Securing primarily privacy
Better Internet Bureau	www.better-internet-bureau.org Canada and other countries all over the world	Meeting a certain level of quality
BizRate	www.bizrate.com Founded by people from all over the world	Internet vendors' rankings

Source: Kjølseth, B. B. 2005. Applying codes of conduct and trust marks. Report from eMarketServices. http://www.emarketservices.com (accessed August, 8, 2007). With permission.

"anonymous," especially on the Web. Chambertrust was designed for the twenty-first century's global 24/7 e-business trade route. Chambertrust is not only an organization but also the worldwide B2B verification mark for trust and confidence in e-business.

Chambersign was founded in 1999 by Eurochambres and by chamber of commerce organizations of 10 European countries (Austria, Germany, Belgium, Spain, the Netherlands, Italy, France, Luxembourg, the United Kingdom, and Sweden). It deals with and discusses widely all aspects of digital signatures appearing in these European countries. It currently covers only the territories of the 10 listed countries, but its further expansion is planned.

Standards for e-business aim at introducing certain standards, norms which could be generally and widely accepted and required in the field of quality. We can distinguish three categories of standards:

- Content—a way of defining the subject of the transaction
- Transaction—these types of standards are still developing
- Communication—nowadays, mostly used standards. (Pahladsingh 2005)

Trust Model in B2B Relationships According to Dyer and Chu

Dyer and Chu's (1997) study examined the determinants of supplier trust in the buyer in automotive industry in the United States, Japan, and Korea. On this basis, they built their own original trust model explaining relationships in B2B. In this model for B2B, trust in supplier–automaker relationships is a sum of the following components:

1. Faith in automaker (manufacturer) that he will behave fair and honest; it is the extent to which the supplier has trust in the manufacturer to be treated fairly and justly.
2. Reputation of "reliable" automaker (manufacturer) in the general business environment (for instance, commitments and keeping promises); it is the extent to which the automaker has a reputation as being trustworthy.
3. Taking excessive advantage of the supplier (an exchange partner) when the opportunity appears; it is the extent to which the supplier perceives that the automaker will take unfair advantage of him (reverse scored). (Dyer and Chu 1997, p. 15)

Dyer and Chu found five key elements that are applied in their trust model. These factors, called multiple scale items, were designed to measure the extent to which the supplier trusted in the automaker. They are as follows:

1. **Length of relationships (LENGTH)** is expressed in the number of years that have passed since the collaboration between supplier and automaker started. To be more specific, this factor not only covers length of collaboration but also its intensity. Thus, we would expect the following: "the longer the duration since the first supplier-buyer transaction, the higher the supplier's trust in the buyer" (Dyer and Chu 1997, p. 6).
2. **Face-to-face communication (FACE)** is operationalized as the annual "man-days" that the supplier and the automaker had face-to-face contact during the certain year the data were collected. This measure covers face-to-face contacts made between automaker purchasing and engineering personnel and supplier sales and engineering personnel. The "man-days" are calculated by identifying the number of personnel from the automaker that directly communicated with the supplier personnel. We can assume that "the greater the face-to-face interaction between the supplier and buyer, the higher the level of supplier trust in the buyer" (Dyer and Chu 1997, p. 7). Both measures, length of relationship and face-to-face communication, belong to the relation-based (embeddedness) variables.
3. **Continuity of the relationship (CONTINUITY)** refers to a model change, which is a natural time for buyers to reevaluate suppliers and implement any changes that are required and needed. Examples of these changes are changes in demand, participation in business projects, and renewing or making some changes in current contracts. It is calculated as the percentage of time the supplier's business had been renewed. The longer history of "re-winning" the business by the supplier, the greater continuity in their relationships with the automakers. A supplier with a low "re-win" percentage is believed to be less trustworthy. To sum up, "supplier trust is higher when the buyer has a track record of continuous repeated exchange with the supplier (e.g., the supplier's history of re-winning the contract at the model change is high)" (Dyer and Chu 1997, p. 9).
4. **Assistance-giving routines (ASSISTANCE)** include three main types of assistance to the supplier: assistance in the field of the transaction subjects'

quality, assistance with cost cutting (e.g., transaction costs, manufacturing costs), and assistance with inventory management. It is essential that all these actions linked with reducing costs and improving quality and inventory management be conducted on a regular basis, not only once or rarely. It is believed that "the greater the assistance provided by the buyer to the supplier, the greater the supplier's trust in the buyer" (Dyer and Chu 1997, p. 10).
   Continuity of the relationship and assistance-giving routines are process-based variables.
5. **Stock ownership (STOCK)** is an economic-based hostage. It is operationalized as the percent of supplier stock (supplies) owned by the automaker. Financial liquidity is crucial for every supplier, and it originates from stock liquidity. With reference to trust in the buyer in the automotive industry (or in general in B2B relationships), we can state that: "the greater the buyer's ownership of supplier stock, the higher the level of supplier trust in the buyer" (Dyer and Chu 1997, p. 11).

Dyer and Chu (1997) carried out research grounded in this model in the United States, Japan, and Korea. The research involved 30 managers in 8 automakers and 70 of their suppliers. The importance of each of the components was measured on a 7-item Likert-type scale (with 1 representing "Not at all" to 7 representing "To a very great extent"). Results were given for a certain period of time, usually a year. In consequence, they formed the basis not only for a static analysis (level of trust in a specific period of time) but also a dynamic analysis (comparison of the results derived from various periods of time). That has enormous meaning for defining the effectiveness of the trust relationship building processes.

Furthermore, the presented trust model can be stated in the form of the following formula, in which Dyer and Chu examined trust in B2B relationships in the automotive industry with a linear regression model:

$$\begin{aligned} &\text{TRUST} = \\ &= a + b_1 * \text{LENGTH} + b_2 * \text{FACE} + b_3 * \text{CONTINUITY} + \\ &+ b_4 * \text{ASSISTANCE} + b_5 * \text{STOCK} \end{aligned} \tag{2.1.2}$$

where a, b_1, b_2, b_3, b_4, b_5... are variables in a linear regression model.

Dyer and Chu described five measures that are significant in building trust within the automotive industry. Therefore, it would be illuminating to analyze the results of their research in the United States, Japan, and Korea. They applied a linear regression model because there was no reason to assume a nonlinear relationship among the estimated variables. It was found that supplier trust is considerably higher in Japan than in Korea or the United States. In the final analysis, Dyer and Chu conclude that:

1. Length of relationship between the supplier and the automaker heavily affected a trust-building process. This length had a meaningful, positive effect on trust in B2B relationships.
2. Face-to-face communication did not influence trust. We can conclude that more important is the communication itself than its form. We do not need only face-to-face communication, but all kinds of communication in order to intensify trusting relationships.
3. Continuity in the relationships was positively associated with supplier trust.
4. Assistance given by the automaker affected positively the level of supplier trust, but only in Japan and Korea. The assistance-giving routines have in these countries very long traditions contrary to the United States, where it is not so common. The higher quality level and the bigger area of assistance offered to the supplier, the higher the trust.
5. Stock ownership did not influence trust.

These findings are a valuable source of information on how to build trusting relationships in B2B. Only three from the initial five variables in their trust model have a significant effect on trust in supplier–automaker relationships. The results indicate that supplier trust is highly correlated with the length of relationships, relationship continuity, and assistance. Moreover, developing trust is always followed by certain costs, for example, assistance given by an automaker requires employing some additional workers (consultants who manage only contacts with suppliers and continuity of the relationships), which might result in rejecting a more beneficial offer from a new automaker in order to gain a high "re-winning" percentage. Indeed, these costs can increase the importance and significance of the trust-building processes. There is a marked need for further research on trust in B2B and B2C relationships in other sectors and branches, which will be a milestone in developing and improving trust building-processes.

REFERENCES

Atena, G. L. 2003. Digital signature. Report from eMarketServices. http://www.emarketservices.com (accessed August 8, 2007).

Bailey, B. P., Gurak L. J., and Konstan, J. A. 2002. Trust in cyberspace. In *Human factors and Web development*, 2nd ed., J. Ratner (ed.), 311–321. Mahwah, NJ: Lawrence Erlbaum Associates.

Dyer, J. H., and Chu, W. 1997. The determinants of inter-firm trust in supplier-automaker relationships in the U.S., Japan, and Korea. http://imvp.mit.edu/papers/98/146a.pdf (accessed August 8, 2007).

E-commerce and development report 2002. 2002. New York and Geneva: United Nations.

eMarketServices. 2004a. eBusiness issues: Building confidence. http://www.emarketservices.com (accessed August 8, 2007).

eMarketServices. 2004b. eMarket basics. http://www.emarketservices.com (accessed August 8, 2007).

EU Commission. 2002. Open consultation on "Trust barriers for B2B emarketplaces": Presentation of the main results. http://ec.europa.eu/enterprise/ict/policy/b2b-consultation/b2b-trust-cons-sum.pdf (accessed August 8, 2007).

Grieger, M. 2003. Electronic marketplaces: A literature review and a call for supply chain management research. *European Journal of Operational Research*, 144, 280–294.

Handfeld, R., and Bechtel, C. 2002. The role of trust and relationship structure in improving supply chain responsiveness. *Industrial Marketing Management*, 31, 367–382.

Hsiao, R. 2003. Technology fears: Distrust and cultural persistence in electronic marketplace adoption. Journal of Strategic Information Systems, 12, 169–199.

Kjølseth, B. B. 2005. Applying codes of conduct and trust marks. Report from eMarketServices. http://www.emarketservices.com (accessed August 8, 2007).

Kjølseth, B. B. 2006. E-business report: Norway. Report from eMarketServices. http://www.emarketservices.com (accessed August 8, 2007).

Laudon, K. C. and Traver, C. G. 2002. *E-commerce: Business, technology, society*. Boston: Addison-Wesley.

McVey, A. 2005. Wine: B2B trading on e-marketplaces. Report from eMarketServices. http://www.emarketservices.com (accessed August 8, 2007).

Pahladsingh, S. 2005. Standards for e-business. Report from eMarketServices. http://www.emarketservices.com (accessed August 8, 2007).

Phillips, C., and Meeker, M. 2000. The B2B Internet report: Collaborative commerce. *Equity Research North America*, April, 23. New York: Morgan Stanley-Dean Witter.

Raisch, W. D. 2001. *The e-marketplace: Strategies for success in B2B e-commerce*. New York: McGraw-Hill.

Schneider, G. 2002. *Electronic commerce*, 3rd ed. Boston: Thomson Course Technology.

Singh, V. 2004. E-marketplace development. Report from eMarketServices. http://www.emarketservices.com (accessed August 8, 2007).

U.S. Electronic Signatures in Global and National Commerce Act of 2000 (ESIGN Act). Public Law 106–229, Section 106. Definitions, 114 STAT 472, June 30, 2000.

Vargas, J. P. 2005. Security and certification systems. Report from eMarketServices. http://www.emarketservices.com (accessed August 8, 2007).

Zällh, S. 2005. Significant e-marketplaces. Report from eMarketServices. http://www.emarketservices.com (accessed August 8, 2007).

3 Virtual Organizations

3.1 THE MULTIFACETED NATURE OF VIRTUAL ORGANIZATIONS

Each era provided its signature form of organization (Lipnack and Stamps 2000, p. 35–36) due the consequence of the evolution of societal skills:

- In the Nomadic Era, beginning between 2 to 3 millions years ago, people mostly bettered their *small-group* skills as hunter-gatherers. At this point in time, they captured the ability to speak, make simple tools, and set up small "organizations" with the next of kin.
- *Hierarchy* with its ranks came up with the remarkable development of agriculture and then feudalism. The agricultural era began approximately 10,000 to 12,000 years ago and bore the mark of a dramatic shift from the Nomadic Era. Cultivation and herding replaced hunting and gathering.
- The Industrial Age with its very considerable number of factories brought *bureaucracy*. Deeply conditioned by ideas and technologies from the fifteenth century, this age became striking in the eighteenth century, dominant in the nineteenth, and eventually mature in the twentieth. The population at that time exploded and urbanized tremendously.
- The Information Age with its clear imperative to connect gave birth to the digital civilization structure, *networks*. The world's leading economies have become knowledge-based, electronically connected, and globally interdependent. Virtual and intelligent organization conceptions are the organizing imperatives of the modern management system of a modern enterprise in the Information Age (Grudzewski and Hejduk 1998, p. 34).

In the big picture, the birth of a new age literally requires the adoption of new organizational forms and collaborations, which differ from traditional ones. Thus, organizations nowadays witness dynamic business environments, where the level of complexity and the pace of change of clients' needs are very high. Kaplan and Norton (1996) associate that a value for a client is quality, price, appropriate time, and functionality. It is therefore not surprising that Christopher (1998) suggests that the prominent role in organizational thinking and gaining competitive advantage is defined by results that might be boiled down to three adjectives: better, cheaper, and faster. The implementation of such a strategy is facilitated by virtualization (Figure 3.1).

Virtualization has become a widespread acknowledged paradigm in management sciences. Several following megatrends are behind the drive for virtual organizations (VOs):

- Products and services are mostly based on information and knowledge. They therefore can be launched, marketed, and delivered virtually in a digitalized form (Skyrme 1998, p. 25).
- The Internet revolution determined the manner in which business is conducted and allows work, and services, and even products to be carried out fast at distances with relatively low costs (Skyrme 1998, p. 25).
- Networking and interdependence are new modes of organizing and collaborating that enable access to various resources, joined with flexibility and responsiveness (Skyrme 1998, p. 25).
- With globalization of markets and resources, companies can more easily sell their products worldwide and obtain "world-class" expertise, regardless of their base of business operations (Skyrme 1998, p. 25).
- Fundamental shifts in new business environments and client demands entail adapting a strategy based on three crucial blocks: low cost, high quality, and being quick on the trigger in terms of fulfilling client needs (Keinänen and Oinas-Kukkonen 2000).

When seen in this light, virtualization provides the key to understanding present economic and technological changes and the ways a business is run. Much of it can be found in literal interpretations of the word *virtual,* which is derived from the Latin word *virtus* (Britannica), which denotes that something "might exist" and is "theoretically possible."

According to Webster's Dictionary Online, *virtual* means:

1. Having the power of acting or of invisible efficacy without the agency of the material or sensible part; potential; energizing.
2. Being in essence or effect, not in fact; as, the virtual presence of a man in his agent or substitute.

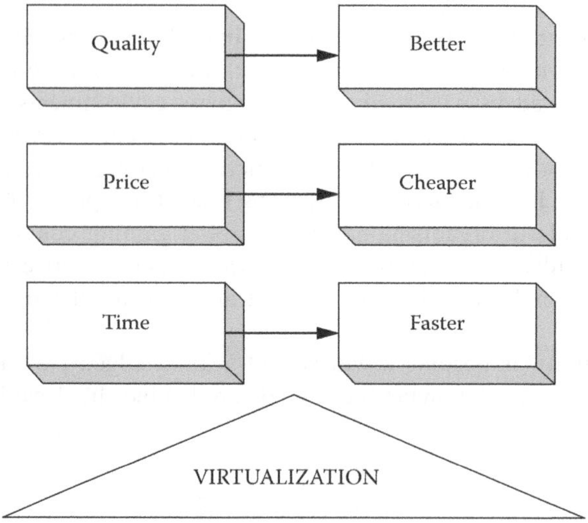

FIGURE 3.1 Creating value for the client.

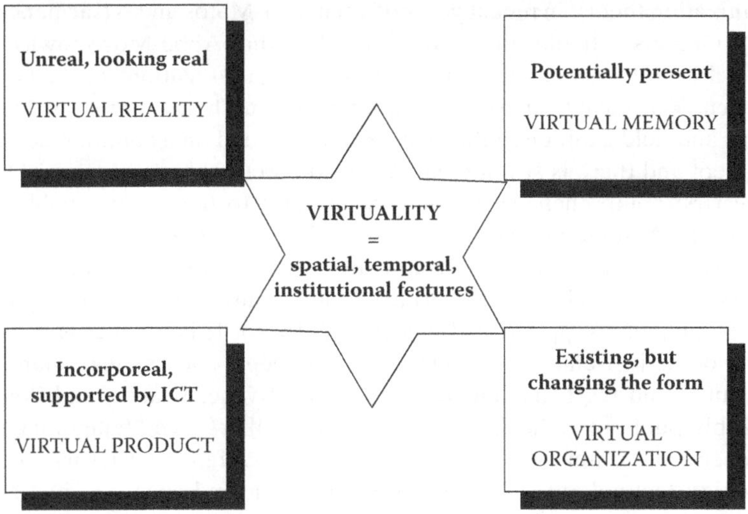

FIGURE 3.2 Concept of virtuality by Bultje and Van Wijk (1998).

The word *virtual* can be interpreted as a force or effect that exists but cannot be directly observed (Grudzewski and Hejduk 2002, p. 41). The philosophical sense of the term *virtual* is boiled down to what is potential, but not tangible, what is an interminable source of a mass of realizations (Jaroszuk 2003). Virtuality is timeless and spaceless. It does not exist physically but does function in reality, as we can observe upshots of its activity. The concept of virtualness is presented in Figure 3.2.

The emergence of the virtualness idea pertains to modern VOs. VOs, which are also recognized by such names as "virtual cooperations" (Davidow and Malone 1992), "virtual enterprises"* (Cheng 1996), and "imaginary organizations" (Hedberg et al. 1997), is said to be the new organizational paradigm.

The origins of this paradigm can be unearthed in the early works of economists like Oliver Williamson in the 1970s. In his prolific body of work, and in particular in the *Markets and Hierarchies*, Williamson undertook an examination of transaction economies** as one of the first and most influential contributions to an economic theory of organization. His ideas had a discernible impact on the "outsourcing wave" in the 1980s (Camarinha-Matos and Afsarmanesh 2005, p. 3–4).

As likely as not, the concept of VO was not invented by a single researcher; rather, it is a theoretical construct that has evolved through a long evolution process

---

* In order to draw a precise picture of the virtual organization concept, one must notice that the terms *virtual organization* and *virtual enterprise* are not exactly interchangeable, although one can come across that in the literature. Virtual organization is a broader concept than virtual enterprise since the latter aims at making a profit; the term *virtual enterprise* does not embrace nonprofit organizations. Notwithstanding this distinction, the term *virtual organization* will be used in the book when meaning a virtual enterprise because it is more contained.

** Transaction costs are costs incurred in the process of an economic exchange. For example, they pertain to searching, information, decision, bargaining, control, handling, and execution costs. The theoretical framework of transaction costs was first discussed by Ronald Coase in 1937.

of organization theory in recent years (Camarinha-Matos and Afsarmanesh 2005, p. 3–4). One also should not overlook the fact that Abbe Mowshowitz is commonly believed to have coined the term "virtual organization" in the late 1980s. Since then, VOs have been given attention by researchers within a wide range of domains and fields, although they have not yet agreed on a common definition of the concept and there is still a great lack of theoretical bedrocks for VOs.

The first comprehensive framework brought to the general public was by Davidow and Malone in 1992 in their book *Virtual Cooperation* (Davidow and Malone 1992). According to Davidow and Malone (1992), virtual corporation "appears almost edgeless, with permeable and continuously changing interface between company, supplier, and customers" (p. 5). It is an ever-varying group of common or just-emerging management concepts such as lean management, just-in-time, and trust management. Later on, Byrne, Brandt, and Port (1993) remarkably publicized the term VO in *Business Week* as a "temporary network of independent companies—suppliers, customers, and even rivals—linked by information technology (IT) to share skills, cost, and access to one another's markets that come together quickly to exploit fast-changing opportunities." In this sense, the virtual business model of collaboration is fluid and flexible, as it refers to a group of voluntary collaborators that quickly unite their capabilities to benefit from a specific fast-changing opportunity. Once the opportunity is met, a VO more often than not is dissolved (Byrne, Brandt, and Port et al. 1993).

Literature has given divergent definitions of VOs and, in many cases, the term VO has been conceptualized in literally dozens of ways. Notwithstanding this fact, generally we can distinguish two distinctly different views on VOs. On the one hand, there is an **institutional** (or structural) view represented by such authors as Davidow and Malone (1992) and Byrne, Brandt, and Port et al. (1993); on the other hand, there is a **functional** (or process) perspective represented, for instance, by Venkatraman and Henderson (1998).

The institutional (structural) approach concentrates on the development and improvement of existing coordination mechanisms for VOs and drawing the distinction between VOs and other organizational forms. A VO is created on the basis of voluntary participation, and its participants engage in various interactions in order to achieve a common objective (Kisielnicki 2002).

A generalized VO definition for the structure-driven approach can be formulated in the following manner: a temporary network of independent collaborators (e.g., companies) usually joined (albeit, not necessarily) by means of information and communication technologies (ICTs) for the purpose of sharing skills, costs, and market success. A network is usually disbanded when the common goal is reached. Ideally, each collaborator contributes to the collaboration only its core competencies. In itself, the network is based on mutual trust. A main object of an institutional approach analysis is examining building blocks of the VO structure and their properties. According to its critics, the institutional approach miscarries in some intrinsic issues, namely:

- Offering information on processes that must be designed and set in place to support this organizational form
- Bringing a brainwave into the differentiation between the processes that endorse one organizational form over another.

Proponents of the **process perspective*** (e.g., Hale and Whitlam 1997; Venkatraman and Henderson 1998; Grudzewski and Hejduk 2002) emphasize in turn the strategic reflection of VOs and the role of ICTs in the processes of continuous changes and improvement of the competitive position. Typical examples of definitions of a VO in accordance with the process approach are as follows:

- "Virtual organization is an organization that adopts the ability of the organization to consistently obtain and coordinate critical competencies through its design of value-adding business processes and governance mechanism involving external and internal constituencies to deliver differential, superior value in the marketplace." (Venkatraman and Henderson 1998, as cited in Keinänen and Oinas-Kukkonen, 2000, p. 142)
- "The virtual organization is the name given to any organization which is continually evolving, redefining, and reinventing itself for practical business purposes." (Hale and Whitlam 1997, p. 3)
- "Virtual organization denotes a set of geographically dispersed (even globally) organizational units representing a joint undertaking that are selected dynamically—using the process criterion—to accomplish specific tasks in time." (Grudzewski and Hejduk 2002, p. 45)

The above-cited function-driven definitions of a VO follow that virtualness is not taken as a separate structure, but as a strategic option bearing upon every single organization and referring its incremental improvements to the existing business architecture. The VO is therefore a "dynamic tool of management" (Grudzewski and Hejduk 2002). As a matter of fact, the definition of VO according to Grudzewski and Hejduk (2002) encompasses both the structural- and process-based perspectives (VO denotes a set of geographically dispersed organizational units).

The functional approach applies to a wide variety of companies. It emphasizes that ICTs enable an effective organization. Venkatraman and Henderson (1998) even posited that a VO cannot function without the significant support of IT. However, this is a questionable assertion. For instance, the research of Appel and colleagues (2004) attests to the preeminence of managerial skills over the employment of IT (an increase in productivity of 8% versus 2%). The significant issue coming from that research to point out here is that use of IT and managerial skills can cause a synergistic effect (an increase in productivity of 20%), which is meaningful for effectiveness. Thus, when formulating strategy, it is pertinent to understand the potential of ICTs and their benefits.

* A *process* is interpreted as "a goal-producing behavior that is composed of events that constitute changes in the structural properties of the system or its environments" (Ackoff 1971, p. 666).

On the top of that, the institutional view misconstrues the role of ICT as a VO enabler; it remarkably ignores the trust that is commonly acknowledged as a prominent facilitator for VO.

The process perspective highlights the significance of knowledge and intellect in value creation. Furthermore, this view moves from structural emphasis to the strategic dimensions introduced by Porter and Fuller (1986). Even some advocators of the process perspective on VOs tend to use the term *virtual organizing* instead of VO in order to emphasize continuity of the emerging processes in an organization with virtual features. More to the point, the word *organizing* also suggests unceasing changing, which is a basic norm for VO creations. Thus, there is good reason behind the wording "virtual organizing."

In the process view, stress is placed on the omnipresence of learning processes within a VO. It can even be stated that a VO on a high level of development is also an intelligent organization. **Intelligent organization** (according to Pedler as cited; Grudzewski and Hejduk 1998, p. 35) is such an organization that facilitates processes of learning of all its members and is adapted to permanent change. In accordance with Garvin's as cited (Grudzewski and Hejduk 1998, p. 35) concept, a VO specializes in creation, formation, and transformation of knowledge as well as in adaptation of its behavior to the environment so that all reflect this newly acquired knowledge. Intelligent organization (such as a particular VO) determines a company's ability to recognize and adjust to an environment and to find the appropriate mode of getting anticipated results (Grudzewski and Hejduk 1998, p. 35). Grudzewski and Hejduk (2000, p. 75–76) claim that an intelligent organization's underpinnings are **competencies,** which are assisted by such key elements as **curiosity, forgiveness, trust,** and **common acting.** It is a self-improving organization leaning on the identification of core competences indispensable for coordinating activities of a new organization.

The question then arises about which approach better reflects the notion of VO, institutional (treating the VO as a new organizational form) or functional (perceiving and interpreting respective dimensions of virtualness and, on the basis of that, delineating the degree of virtualization). The characteristics of the two discussed approaches are summarized in Table 3.1. We consider that they serve two different purposes. Basically, the structural view is appropriate when researchers want to compare organizations with different degrees of adoption of VO concepts and performance, whereas the process view gains when it comes to implementing programs and design of a distinct VO. However, we must stress that the functional view should also regard the trust component in the design of the VO.

As of today, the term *virtual organization* has been loosely defined in the academic and practical literature. Thus, the term *virtual organization* became a buzzword. There is no commonly acknowledged definition of VO and, therefore, related problems are immediately apparent; specifically, the literature on VOs is limited and mostly conceptual. Consistently, empirical findings are not comparable. Taking those two concerns together, it is obvious that understanding of VOs is still at an embryonic stage and there is still much to learn.

We posit that researchers must now adapt to a certain view of VOs with precisely stated areas of implications and restrictions in order to adequately compare

TABLE 3.1
Characteristics of Two Different Views of VOs

Institutional (Structural) View	Functional (Process) View
Information technology	Information technology
Core competences	Core competences
Flexibility	Flexibility
Blurred boundaries	Blurred boundaries
Flat hierarchies	Attribute of every organization
Temporality	Value-adding business processes
Network of legally independent units	Learning and adaptive orientation
Common business purpose	Usually intraorganizational perspective
Based on trust	Useful for implementation programs
Usually interorganizational perspective	Venkatraman and Henderson (1998), Hale and Whitlaw (1997)
Useful for a research insight	
Byrne (1993), Davidow and Malone (1992), Goldman et al. (1995)	

Source: Keinänen, K., and Oinas-Kukkonen, H. 2001. Virtual organizing as a strategic approach to stay competitive: A conceptual analysis and case study. In *Knowledge Management and Business Model Innovation*, Y. Malhotra (ed.), Hershey, PA: Idea Group Publishing. With permission.

empirical findings, with the stress on the range of cases they concern. So, in our view, a preliminary but necessary step is to define what a VO is in order to investigate this phenomenon. For example, here, we have come up with a proposition for a comprehensive definition of VO that concerns **commercial companies' collaboration in an interorganizational context.** A limitation to that could be justified for the purpose of building a solid empirical base useful for business implications on which is contingent the quality and generalization of empirical findings.

Proposition of Definition 3.1 Virtual organization refers to a changing (according to switching principle) network of collaborating companies that pursue a common purpose: to harness a fast-changing market opportunity. (Sankowska and Wańtuchowicz 2006)

It is worth quoting the above definition in other words, more descriptive at length:

Definition 3.1 Virtual organization is a name given to any form of a **temporary,** loosely coupled network of [relatively] **independent** companies chosen dynamically (named *collaborators*), often located in different places, that are brought together to fulfill a **common purpose**—to harness a fast-changing market opportunity—developing, producing, and/or offering the appropriate product(s) and/or services. It is based more on trust than formal contracts. Once the specific goal is met, the venture will, more often than not, disband. Each collaborator brings to the cooperation its **core competencies**. (Sankowska and Wańtuchowicz 2006)

In order to get a more comprehensive view on VOs, it is very important to be acquainted with the constituents of an ideal VO. Such a VO does not exist, but it is a good starting point for the analysis and estimation of a degree of adoption of real organization toward the ideal one. So, a binary classification of companies (virtual versus nonvirtual) can be replaced by a concept of **gradual virtualization**, which has been so far very well received by researchers. In the following, we list and briefly describe the most important characteristics of the ideal VO that can be exhibited by a real organization to some extent. They are as follows:

- **Focus on core competences* of the VO's members**—the selection of members according to their skills, knowledge, rare resources (competences) in order to form a purposeful, world-class, excellent organization
- **Networking**—the VO constitutes a network of legally independent, different collaborators that unite to accomplish a certain task
- **Flat structures**—the VO does not need to have a hierarchy but does have a high degree of decentralization
- **Fuzziness of boundaries**—organizing principles of the VO often make it difficult to discern where each organization begins and ends due to the fact that the VO is a combination of independent organizations managed as if they were one entity
- **Temporariness**—the VO exists as long as it serves the specific purpose for which it was created
- **Flexibility**—the VO responds actively and quickly to changes
- **Law degree of formalization**—relationships between partners in the VO are less formal and less permanent; these relationships create links, but a partner can still survive without the other partners as they preserve their independence
- **Common business purpose**—the VO appears in order to pursue a common business goal
- **Sharing risk, resources, and knowledge**—skills, knowledge, and costs related to task completion, risk of failure, and access to global markets (positive and negative aspects of cooperation) are shared by partners in the VO
- **Trust**—the VO, which hardly ever uses legal contracts, requires a higher amount of trust between participants than ever; under the circumstances, the VO should be guided by the "win–win" approach to others
- **Time and spatial dispersion (globalization)**—often, components of the VO are dispersed in time and space and the VO exceeds frontiers; fiscal distance between participants is believed to stop to play the role**

* Hamel and Prahalad (1999) denoted that core competencies are key traits of an enterprise's competitiveness that contribute significantly to the value perceived by a client. They refer to unique knowledge and are the source of access to new markets.

** A note on that fact: There have come into sight new obstacles in performing work across space as the result of the "face-to-face" communication reduction. This implies that socio-emotional aspects of work at distance are still brain teasers.

- **ICTs**—the VO uses ICTs, which mainly facilitate communication despite diverse locations
- **Customer based**—the VO endeavors to satisfy perfectly the needs and wishes of its customers
- **Modularity**—the VO is based on small, customer-orientated processes composed of relatively small, manageable units called *modules*
- **Heterogeneity**—participants of the VO possess different core competencies with regard to assigned tasks
- **Switching**—refers to the fact that a VO alternates its composition over time and exploits the advantage of switching accomplished through a dynamic assignment (Mowshowitz 1994, 2002)
- **One virtual identity**—the VO is a set of independent units but is considered from the viewpoint of the outsider (by a final client) to constitute one common organization
- **Market-opportunity orientated**—companies come together to take advantage of a fast-changing market opportunity and, more often than not, disengage when the specific opportunity is exploited or when demand ceases to exist.

Referring to the most significant features of the ideal VO, one can investigate the degree of their adoption for a specific organization. Consistently, one can classify a considered organization according to its degree of virtualization. This allows researchers to draw a profile of virtualization applying previously described definite dimensions.

Manifest features of a VO are given a great deal of attention by researchers. However, they have not yet reached concordance on the matter of their explicit set. The selection of dimensions from the list of hypothetical VOs is still debatable. Some authors acknowledge that certain dimensions (e.g., temporariness, spatial dispersion) are not necessary conditions for organizations to exist as VOs.

For moving forward to a better understanding of the VO idea, there is presented an integrated model of VO conceived by Saabeel, Verduijn, and Kumar (2002). The model combines the structure and process perspectives of VOs and consists of three interdependent layers: the universe of modules, the dynamic web, and the dynamic organization. The descriptive process approach in the model of Saabeel and colleagues is in fact reduced to adumbrating the phenomenon that acts as a trigger for the form change. Each layer in this model represents a structure constituent (structural approach) and transitions between layers (representing process aspects/functional approach) (Saabeel et al. 2002, p. 9). According to the three-layered model of organizations that formulate a VO, dynamic organization is a part of a larger network—the dynamic web (virtual web)—of which a final selection for the dynamic organization is made. It implies that organizations forming a VO at a particular moment are already no stranger to each other (i.e., they know each other from previous interactions or they have some information on themselves). Goldman, Nagel, and Preiss (1995) introduced the term **virtual web** as an open-ended collection of pre-qualified partners that agree to form a gathering of potential participants of VOs. The **universe of modules** is the set of all organizations, and it pulls together the individual organizations from, for instance, a specific industry or economy. Whereas in VOs

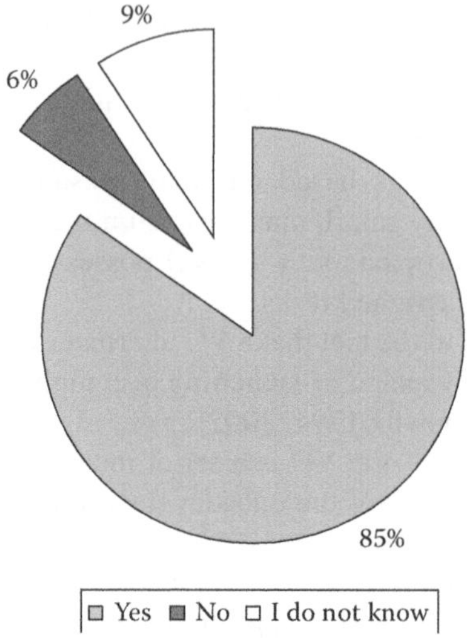

GRAPH 3.1 Companies were asked if they envisaged that VO employment will be the source of a competitive advantage between now and 2010? (n = 33). This graph represents the response. *Source:* Sankowska, A., and Wańtuchowicz, M. 2006. *Organizacja wirtualna jako forma organizacji sprzyjająca rozwojowi przedsiębiorstwa.* Unpublished Statutory Works, Faculty of Production Engineering, Warsaw University of Technology, Warszawa. With permission.

the main objective is to cash in on a specific market opportunity, the purpose of the **dynamic web** is a constant disposition to work together in the future when market opportunity occurs.

Market opportunity or demand is the initiator that pulls organizations together. A market opportunity delineates a (latent) area of need or requirement in which a company can turn a profit (Kotler 1994). The VO can fulfill this need by brining together the complementary core competencies required for developing, producing, and offering the appropriate products and/or services (Saabeel et al. p. 9). A **market demand** may be defined as a concrete desire for specific products that are established as effectual both by an ability and a willingness to buy them (Kotler 1994).

In spite of this extensive literature discourse on VOs, there is a question about its practical implications. Unsurprisingly, 85% of Polish companies (n = 33) that apply the concept of VO in the sense of Definition 3.1 believe that between 2006 and 2010, VO employment will be the source of a competitive advantage (Graph 3.1).

While a number of authors posited that VOs are an effective mode of organizing, at present, there seems to be little or no comprehensive literature available that either empirically justifies this assumption or ascertains logical arguments for it. A number of issues related to this still await both consideration and clarification. The initiation of the VO makes sense until it is justifiable. We can look for some justification in some pilot studies. The VENICE Project surveyed 12 virtual enterprises and commenced in May 1998. The results of the investigation revealed an overall strong enthusiasm for the organizational form of VO among enterprises. The top seven perceived benefits of VOs chosen by respondents are operational flexibility,

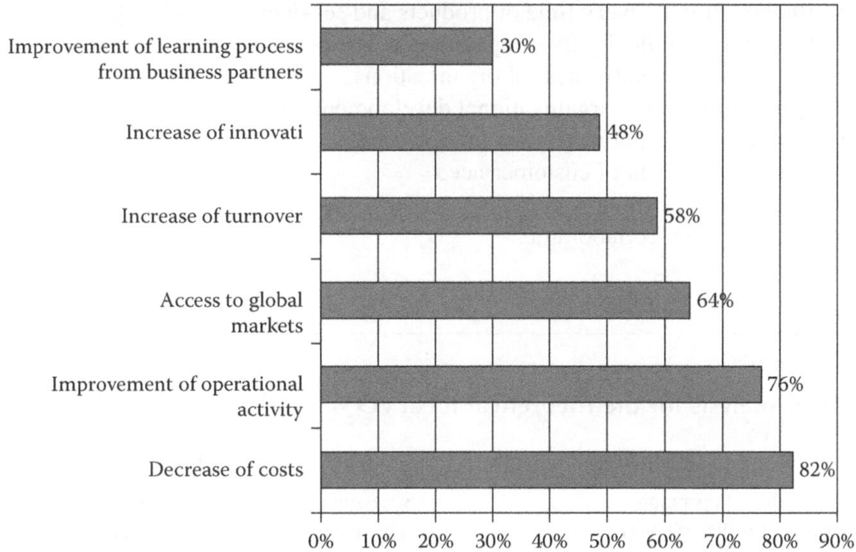

GRAPH 3.2 Perceived benefits for operating as a VO (n = 33). Please note that answers do not add up to 100% because respondents were able to give multiple answers to this question. *Source:* Sankowska, A., and Wańtuchowicz, M. 2006. *Organizacja wirtualna jako forma organizacji sprzyjająca rozwojowi przedsiębiorstwa.* Unpublished Statutory Works, Faculty of Production Engineering, Warsaw University of Technology, Warszawa. With permission.

productivity, cost savings, quick action (rapid response to new opportunities), access to global markets, ability to network, and exploitation of core skills. Comparable results to that brought the coauthors research (Graph 3.2).

The most common disadvantages and drawbacks of VO systems are problems in communication concerning technical issues as well as the clarity of conveyed information. A potential source of problems within VOs is also an occurrence of unclear (blurred) responsibilities and the phenomenon of social isolation, which entails the necessity of face-to-face meetings. Seeing that even one informal (direct) meeting can promote establishment of social bonds and therefore increase the likelihood of success, situations in which partners never meet in the real world are rather rare.

To better crystallize the picture of VO, one can evaluate strengths, weaknesses, opportunities, and threats involved with the ideal (hypothetical) VO by SWOT analysis. In Table 3.2 are listed the results of this technique, pertaining to both the attributes of VOs and external conditions that affect the achievements of VO objectives. One of the most self-evident disadvantages of a VO is that one casts doubts on entities, as often they cannot be situated in the "real world." Thus, it is necessary to build some suitable guarantees and protection mechanisms (Grudzewski and Hejduk 2002, p. 55).

To repeat, VO can be considered too as a strategic option. There are the following strategic reasons to implement this form in practice:

- Risk, cost, infrastructure, research and development activities sharing
- Combing complementary outstanding core competences for the purpose of delivering word-class or just cheap products and services

- Reduction in delivery time of products and services
- Ensuring high flexibility of operating activities
- Improving innovativeness of organizations
- Quickening of the organizational development pace
- Cost-cutting of outlays on the infrastructure
- Better fulfillment of customer need
- Gaining access to new markets
- Leaning from collaborators.

TABLE 3.2
SWOT Analysis for the Theoretical Ideal VO

Strengths	Weaknesses
Big flexibility of operation Big speed of realizing transactions Common operating policy in the range of organizational activities Reducing the cost of realizing transactions Reducing investment outlays on organization development Minimizing legal services associated with transactions Contribution by each partner in the area where he or she is fully professional to a VO	Necessity to have IT, which makes transaction possible to realize, including a) global networks, b) big databases Necessity to trust all the organizations operating within a VO Possibility of including an incompetent and/or unverified organization Difficulties in enforcement of payments for unfulfilled transactions Lack of formal control and the possibility of difficulty in coordinating completion of transaction
Opportunities	**Threats**
Quick reaction to environmental changes, in particular, the appearance of so-called "niches" Realizing transaction despite legal and organizational barriers Possibility to use the most modern, cutting-edge methods and management techniques Possibility of cooperation of such partners who in another organizational model would have never worked together Within informational links joining cooperating partners there are no state duty borders Reaching new groups of clients Absence of racial, cultural, and other types of prejudice in contacts between transaction parties	Fallibility of technical devices facilitating networking Lack of legal regulations for operating of organizations included in a VO group and their responsibility toward each other and their clients Lack of preparation of both clients and organizations to use virtual organizations Necessity to overcome reluctance to changes in the transaction system

Source: Kisielnicki, J. 1997. Wirtualna organizacja jako wytwór ery informacyjnego społeczeństwa. *Organizacja i Kierowanie,* 4(90), 31. With permission.

REFERENCES

Ackoff, R. L. 1971. Towards a system of systems concepts. *Management Science,* 17(11), 661–667.

Appel, A. M., Dhadwal, A., Dorgan, S. J., and Dowdy, J. 2004. *When IT creates value.* http://www.mckinsey.com/clientservice/bto/pointofview/itvalue.asp (accessed January 20, 2006).

Bultje, R., and Van Wijk, J. 1998. Taxonomy of virtual organisations, based on definitions, characteristics, and typology. *VoNet: The Newsletter,* 2(3), 9.

Byrne, J. A., Brandt, R., and Port, O. 1993. The virtual corporation. *Business Week,* 8 February, 36–40.

Camarinha-Matos, L. M., and Afsarmanesh, H. 2005. Brief historical perspective for virtual organizations. In *Virtual Organizations: Systems and Practices,* L. M. Camarinha-Matos, H. Afsarmanesh, and M. Ollus (eds.), New York: Springer.

Cheng, W. 1996. The virtual enterprise: Beyond time, place, and form. *Economic Bulletin,* Singapore International Chamber of Commerce, 5–7 February.

Christopher, M. 1998. *Logistics and supply chain management: Strategies for reducing cost and improving Service.* London: Pitman Financial Times.

Davidow, W. H., and Malone, M. S. 1992. *The virtual corporation: Structuring and revitalizing the corporation for the 21st century.* New York: HarperBusiness.

Goldman, S. L., Nagel, R. N., and Preiss, K. 1995. *Agile competitors and virtual organizations: Strategies for enriching the customer.* New York: Van Nostrand Reinhold.

Grudzewski, W. M., and Hejduk, I. K. 1998. Kreowanie przedsiębiorstwa przyszłości z wykorzystaniem teorii i praktyki organizacji inteligentnej. In *Menadżer u progu XXI wieku,* S. Borkowska and P. Bohdziewicz (eds.). Łódź: WSH-E.

Grudzewski, W. M., and Hejduk, I. K. 2000. *Przedsiębiorstwo przyszłości.* Warsaw: Difin.

Grudzewski, W. M. and Hejduk, I. K. 2002. *Przedsiębiorstwo wirtualne.* Warsaw: Difin.

Hale, R., and Whitlam, P. 1997. *Towards the virtual organization.* London: McGraw-Hill.

Hamel, G., and Prahalad, C. K. 1999. *Przewaga konkurencyjna jutra.* Warsaw: Business Press.

Hedberg, B., Yakhlef A., and Baumard, P. 1997. *Virtual organizations and beyond: Discover imaginary systems.* Chichester,UK: John Wiley and Sons, Ltd.

Jaroszuk, M. 2003. Wirtualność a wartość przedsiębiorstwa. In *Value 2003: Wpływ zasobów na wartość firmy,* Vol. 1, E. Skrzypek (ed.), Lublin, Poland: UMCS.

Kaplan, R. S., and Norton, D. P. 1996. *The balanced scorecard.* Boston: Harvard Business School Press.

Keinänen, K., and Oinas-Kukkonen, H. 2000. Virtual organizing as a strategic approach to stay competitive: A conceptual analysis and case study. In *Knowledge management and virtual organizations,* Y. Malhotra (ed.). Hershey, PA: Idea Group Publishing.

Kisielnicki, J. 1997. Wirtualna organizacja jako wytwór ery informacyjnego społeczeństwa, *Organizacja i Kierowanie,* 4(90), 23–34.

Kisielnicki, J. 2002. Virtual organization as a chance for enterprise development. In *Modern organizations in virtual communities,* J. Kisielnicki (ed.), 100-115, Hershey, PA: IRM Press.

Kotler, P. 1994. *Marketing management: Analysis, planning, implementation, and control.* Englewood Cliffs: NJ: Prentice-Hall International.

Leimeister, J. M., Weigle, J., and Krcmar, H. 2001. Efficiency of virtual organisations: The case of AGI. *Electronic Journal of Organizational Virtualness,* 3(3), 13–43.

Lipnack, J., and Stamps, J. 2000. *Virtual teams.* second edition. New York: John Wiley and Sons.

Mowshowitz, A. 1994. Virtual organization: A vision of management in the Information Age. *Information Society,* 10(4), 267–288.

Mowshowitz, A. 2002. *Virtual organization: Toward a theory of societal transformation stimulated by information technology.* Westport, CT: Quorum Books.

Porter, M. E., and Fuller, M. B. 1986. *Competition in global industries: Coalitions and global strategy.* Boston: Harvard Business School Press.

Saabeel, W., Verduijn, T. M., and Kumar, K. 2002. A model of virtual organisation: A structure and process perspective. *Electronic Journal of Organizational Virtualness,* 4(1).

Sankowska, A., and Wańtuchowicz, M. 2006. *Organizacja wirtualna jako forma organizacji sprzyjająca rozwojowi przedsiębiorstwa.* Unpublished Statutory Works, Faculty of Production Engineering, the Warsaw University of Technology, Warsaw.

Severwright, J. M., and Raghbir, S. VE Technological Requirements: The VENICE Project, Dept. of Computer Science, University College London, UK. www.cs.tcd.ie/Virtues/ocve98/proceedings/006.html (accessed October 20, 2006).

Skyrme, D. J. 1998. The realities of virtuality. *Organizational virtualness: Proceedings of the first VONet workshop,* April 27–28. Bern: Simova Verlag.

Venkatraman, N., and Henderson, J. C. 1998. Real strategies for virtual organizing. *Sloan Management Review,* 40(1), 33–48.

Webster's Dictionary On-line. http://www.webster-dictionary.net/ (accessed November 12, 2006).

Williamson, O. E. 1975. *Markets and Hierarchies: Analysis and Antitrust Implications.* New York: Free Press.

3.2 MODELS OF VIRTUAL ORGANIZATIONS

All constituents of the ideal VO seldom or perhaps never comply with one organization in business practices. For this reason, when speaking about VOs, we often invoke its concrete shape. VOs can erect different forms. Thus, literature distinguishes different types of VOs. Consequently, this is reflected in typologies presented by various authors. In this section, there is provided an overview of the most popular kinds of VOs.

The information-based (i-based) taxonomy of VO suggested by Lethbridge (2001) is an extension of the model created by other authors, such as Burn and Ash (1999). Lethbridge (2001) identified three following broad groups of information flow between members of a VO:

- **Planning (P):** Information used for initiating the common purpose, for setting the background for overall VO activities
- **Operational (O):** Information used to uphold the daily operation of each constituent member
- **Coordination (C):** Information to guarantee that operational activities effectively support the fulfillment of the shared purpose.

The structures of Lethbridge's taxonomy are virtual face, star alliance, market alliance, co-alliance, value alliance, and parallel alliance. They are outlined below.

Virtual Face

Virtual face is the easiest and most well-known VO shape. It embodies basic business-to-customer (B2C) e-commerce or e-business. As a matter of fact, it is not a VO in the strict sense of the word. However, the public most commonly links VO e-commerce. This simple structure can be tracked by its prominent characteristics, namely:

- The e-business form may crop up from the already-existing "physical" business group. In this case, a new extra structure of VO for business is selected. It is simply an enhancement of the ways that a company already operates.

- The core member, as it is only one here, is irreplaceable. It is in every respect responsible for planning, operation, coordination, and customer contact in the digitalized marketplace.
- As there is only one member of such a form, there is no information flow "between members." Therefore, there appears only operational interchange of information with the customer (e.g., orders and delivery notices).

Star Alliance

As soon as a company opts to outsource some of its operations, it can commence running its business as a "star alliance." The leading organization (the core organization) is here in a prevailing position to cope with the customer relationship and comprehensive project management. The core organization is vital to the star alliance and cannot be replaced, for it has unique competences and roles. In contrast, other "satellite" VO members executing a particular function within the VO are characterized by a high substitution rate and can be superseded by a new participant fulfilling the same required function. This VO structure can cover, for instance, a global company. The dominating organization, often operating worldwide, maintains a network of relations with territorially dispersed organizations of a local area of activity (e.g., Reuters).

Activities' planning for this type of VO is done by the main organization. Members of satellites create their own internal plans in agreement with the superior VO plans delivered by the core organization. In all instances, the core organization contacts customers. Thereby, there are only operational and coordination information flows between members. The core organization enjoins on "satellites" regarding to day-to-day activities (operational information) and makes sure that the common goal of the VO is reached satisfactorily with a high standard (coordination).

Market Alliance

A market alliance is established for a concrete purpose: the creation of a common market. Thus, companies (e.g., wholesalers) deliver their products to a shared market for sale and, in return, pay commission for the use of it. However, a group of manufacturers can also establish their own market with a closed number of members, settled up for by subscription and commission fees. The core organization does not manage the market alliance as the market alliance is regulated according to rules agreed upon at the beginning. Whichever the form the market alliance takes, its marketing competencies (e.g., customer relations management) make it irreplaceable. Whereas a satellite can accept rules and join the VO or do aversely, each satellite brings a similar function of goods or services produced for sale. Planning for the VO is or is not worked out before the VO is created which is why only operational information is likely to flow between members. If there are faults in the original plans when they are created, then there is no corrective mechanism in this structure. For the satellites, this implies that whenever the core organization revamps the way it runs its management function, the satellite has no course of action other than to accept these changes or leave the alliance.

ICTs are a pivotal factor in determining market alliance creation, embracing all members first and foremost those dispersed geographically. ICTs connect demand and supply. Organizations coupled with a VO potentially benefit from easier access to new markets and the cost reduction of physical location and distribution. A market alliance enables its members to operate in global markets without concern for geographic boundaries, whereas this alliance offers its clients a diverse choice from the wide range of products and services provided.

Co-alliance

A group of organizations may also agree to work together on a concrete project. Here, there is no core organization, no strong leader; each member of this VO is an equal partner and preserves its independence. Each collaborator brings to the co-alliance a comparative contribution in connection with resources, knowledge, and skills. Each partner fulfills a specific undertaking in the VO and is involved in the planning and coordination process. It may be very difficult to replace any member, particularly within the short life of a single project.

All three categories (P, O, C) of information flow may happen between organization members. Each partner may contact the customer in order to fulfill its assignment in a VO. If one among them undertakes the responsibility to manage the project, then we start to consider the previously described star alliance.

Value Alliance

Here, an organization member receives value from another member in a value chain an output called a "work in progress" (WIP), which is a partially finished product or service that still needs enrichment. Each VO participant then adds some value to the product and advances it closer to completion (seen by the final client) and then transfers the WIP to the next member in the value chain. Information flow is only operational. The WIP is shifted from the one member to another in the value chain. As a result, contact with the client is limited to the first and last organization in the established value chain.

Parallel Alliance

In this VO structure, there is no core organization. Both operational and coordination information flows might be observed. VO members are replaceable, although the necessity for maintaining the coordination and continuity of business make it more difficult to switch a member in a parallel alliance than in a value alliance. Once a parallel alliance is formed, there is no inherent planning. If deviations occur, corrective actions can be undertaken only within the scope set by the inbuilt planning. Some (or all) members of the VO reach clients. By what each member of the VO obtains and provides information connected with its particular task in the VO.

At this point, it is informative to bring in another VO model, the **virtual broker** proposed by Burn and Ash (2000), which is a complement to the above-mentioned set of VO structures. A virtual broker, sometimes also called a net-broker (Miles and Snow 1986), is a designer of dynamic networks. He seeks market opportunities either

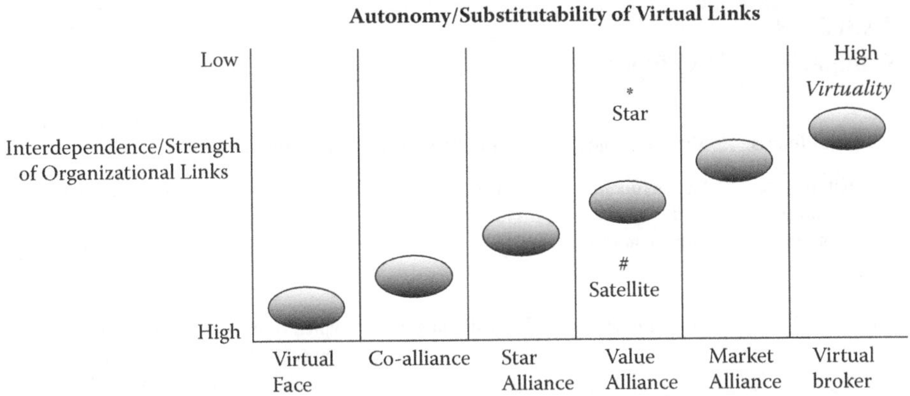

FIGURE 3.3 Virtual alliance models. *Source:* Burn, J.M., and Ash, C. 2000. Knowledge management strategies for virtual organisations. *Information Resources Management Journal*, vol. 13, no. 1, pp. 15–23. With permission.

as a part of an already-existing supplier or as an information broker who belongs to a separate organization specializing in offering specific services in the area of business information. He is responsible for selecting suitable partners, overall project management, and supporting the collective learning processes. The virtual broker is the most temporary kind of VO, which focuses on taking advantage of occurring opportunities.

Each of the presented forms of collaboration has a different virtual degree, denoting both the level of autonomy and interdependence within an organization as well as of flexibility in terms of creating the VO for exploiting new opportunities and for dissolving it when a project is finalized. Figure 3.3 is a graphical representation of the above-presented outline of different VO structures with regard to the degree of virtuality, interdependence of organizational links, and autonomy of virtual links.

Palmer and Speier (1997) developed a different typology of VOs composed of four specific concepts: virtual teams, virtual projects, temporary VOs, and permanent VOs. Table 3.3 gives a comparison.

The term *virtual organization* is loosely defined in the literature. For example, McDonough, Kahn, and Barczak (2001) describe a **virtual team** in these words: "comprised of individuals who have a moderate level of physical proximity and are culturally similar." To distinguish virtual teams from global teams, McDonough and colleagues include in their definition of *virtual team* moderate physical proximity. The key characteristic of virtual teams is that they work predominantly through ICTs but not necessarily in geographically disperse locations. For virtual teams, it implies that their members may also be situated in proximate offices in the same building.

The idea of a **virtual project** is based on organizations that form alliances or consortiums by joining together compatible enterprises with the aim of making good use of an occurred market opportunity. It is commonplace for partnering members to come from similar industries or sectors.

TABLE 3.3
Comparison of VO Types

Virtual Teams	Virtual Projects	Temporary VOs	Permanent VOs	Communities of Practice
Internal to an organizational function or departmental unit	Across functions and organizations	Across organizations	Across organizations	Across functions and organizations
Small, local	Indeterminate	Typically larger	Typically smaller, but scalable	Depending on expertise knowledge
Teams on specific, ongoing tasks	Multiple organizational representatives working on specific projects	Multiple functions responding to a market opportunity	All function and full functionality as a working organization	Knowledge exchange
Membership varies, but form is permanent	Temporary	Temporary	Permanent	Permanent
Connectivity, sparing embedded knowledge (e-mail, groupware)	Repository of shared data (databases, group)	Shared infrastructure (groupware, WANs)	Channel for marketing and distribution, replacing physical infrastructure (Web, Intranet)	Connectivity
Medium	Medium	High, but hard to achieve	High	Medium

Temporary VOs are an extension of the virtual project design, built to engage in multiple projects and to perfect a quick response to a specific market occasion. When a specific market opportunity becomes exhausted, the VO is dismissed.

Permanent VOs represent the modus vivendi organization (Hale and Whitlam 1997) that from its genesis is designed to bind together market players and respond to favorable occasions for both revenue generation as well as cost savings (Palmer and Speier 1997).

In addition to these four types of VOs identified by Palmer and Speier (1997), some authors (e.g., Skyrme 1998) recommend an inclusion of **telework.** Telework pertains to telecommuting, when an employee is working away from home temporarily or permanently and commuting is replaced by ICTs (Nilles 2003). However, an in-depth look at the notion of telework leads to a significant conclusion. Distinct from a teleworker, a virtual team member always works as a part of team, not independently. Thus, telework may or may not evolve into a VO. To put it simply, virtual team and telework are not interchangeable in terms of VOs as a form of cooperation. This is only one of many examples of the misuse of terms with respect to VOs. More precisely, one can postulate that even virtual teams can be differentiated from VOs. In a sense, the VO is the phenomenon of a collaboration that might be organized with the use of virtual teams and it is superior to the virtual team concept. Moreover, a virtual team may or may not be composed of external members from other organizations.

Wenger (1998) added **communities of practice** to the VO types introduced by Palmer and Speier, which spring from the idea of knowledge sharing. Communities of practice are understood as groups of people who meet together (face-to face or electronically) in order to share knowledge. Professional interests, the need for collective problem solving, sharing experience, and so forth are reasons that bring them together (Fazlagić 2004, p. 77). Here, we must point out that classifying a community of practice as an interorganizational VO in the strict sense of collaboration between organizations can be justifiable if we refer to organizing members from different organizations and not within a specific one.

Another classification of VO, which has been widely adopted by researchers, was developed by Bultje and van Wijk (1998). They broke VOs into four groups: internal VO, stable VO, dynamic VO, and web-company.

Internal VO

An internal VO can be interpreted as an enterprise composed of a few scattered business units in the form of autonomous various teams and groups. Managerial actions are executed in a decentralized manner. The high availability of employees based on the implementation of virtual teams and telework is the prime factor for the flexibility of this organization structure.

Stable VO

The main idea behind this concept is the cooperation between different committed organizations of suppliers that are closely attached to a core organization. A leading partner contracts nonkey competencies out to other members. Examples of such organizations include Nike, ING, and Airbus.

Dynamic VO

A dynamic VO is literally a reply to very dynamic, continuously changing market opportunities based on a large cooperation between companies. It is defined as temporal, even ephemeral. The differentiating characteristics of a dynamic VO from a stable VO is that the latter is a more permanent configuration in terms of its members' composition. An example of a dynamic VO is TMG.

Web-Company

A Web-company is a nonpermanent network of specialized organizations operating on the basis of the use of the Internet. In this case, the Internet is the facilitator of products or services offered. The chief role for achieving satisfactory performance is knowledge management. The most recognizable case of a Web-company is Amazon.com.

All characteristics of an ideal VO provided in the section "The Multifaceted Nature of Virtual Organizations" are highly unlikely to be observed in one organization. As a consequence, there were elaborated evolutionary models of virtuality

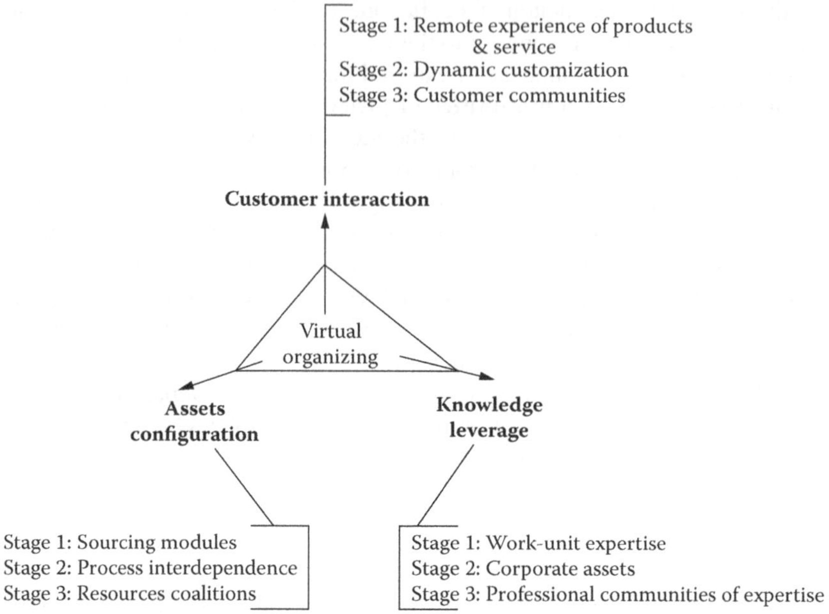

FIGURE 3.4 Model of virtual organizing by Venkatraman and Henderson (1998).

of feature. The literature on virtual organizing notably focuses on two models: Venkatraman and Henderson's (1998) descriptive evolutionary model of virtuality and Scholz's (2000) three-dimensional virtual cube model. They are developed on the same assumption that a VO is a strategic characteristic applicable to every organization that is more an intraorganizational perspective than an interorganizational perspective. Both models are reported below.

Venkatraman and Henderson's **descriptive evolutionary model** emphasizes the key role of ICTs in rending virtual organizing processes. By virtualization, Venkatraman and Henderson (1998) mean an organization's ability to continuously attain and control the core competences through both the processes purposed for the value-added creation as well as internal and external managerial mechanisms targeted at delivering unique, desirable value on the market. The model sprang up from extensive researched carried out by Henderson and Venkatraman, starting as early as 1991, and the strategic dimensions invoked by Porter and Fuller's (1986) intelligible model of value chain. Venkatraman and Henderson's theory discerns three virtual dimensions of a company—**customer interaction** (virtual encounter), **assets configuration** (virtual sourcing), and **knowledge leverage** (virtual expertise)—which are shown in Figure 3.4. Each of these three virtual dimensions encompasses three stages.* The marriage of these dimensions and stages determines virtualness.

The first dimension of virtualness, customer interaction, pertains to new ideas and opportunities coming from company to customer relations. It involves three following stages: remote experience of products and services, dynamic customization

* The former model by Venkatraman (1995) also covers three stages, although they were the following: market experience, competence leverage, and work configuration.

of products and services, and customer communities. **Remote experience of products and services** plays a prominent role in processes of innovativeness that are customer driven. In other words, it indicates the shift from the old product-centric market-share approach to a new customer-centric and profit-centric approach. That follows the need of creating the superior linkages with clients and establishment of manners in which products/services - or cost - driven can be experienced remotely.

Venkatraman and Henderson's concept of **dynamic customization of products and services** is deeply embedded in three ground rules: modularity, intelligence, and organization:

- *Modularity* means organizing complex products/processes/services by organizing smaller flexible systems, so-called *modules,* so that an increased rate of innovation, cost reduction, and above-all greater client satisfaction are gained.
- *Intelligence* breaks loose from permanent current information exchanged with customers which is simultaneously exerted to match up homogenous interest profiles and make recommendations on the basis of customers' shared interests.
- *Dynamic customization of products and services* touches on an organization that operates in a distinct, original manner. Venkatraman and Henderson (1998) stress that engagement in modularity and intelligence would be wasted without adjustment of relevant organizational forms. (Keinänen and Oinas-Kukkonen 2001, p. 147)

Customer interests communities transfer economic force from manufacturers to customers. Prahalad and Ramaswamy (2004), in their book entitled *The Future of Competition: Co-creating Unique Value with Customers*, advocate the essential role of the client in the process of making innovations in terms of searching and evaluating the ideas of new products and services. Analogically, Venkatraman and Henderson (1998) state that the most profound aspect of the virtual model of interaction is the emergence of customer communities groups of common interests, which enable gathering, creating, and spreading valuable information.

The second dimension, **assets configuration,** aims at orchestration of critical assets and resources and is based on the three following stages, in order of their growing advancement: **sourcing module, process interdependence,** and **resource coalitions.** Venkatraman and Henderson (1998) vouched for constant strategic inspection which sourcing modules can be acquired from external sources and how existing sourcing logic should be reorganized and improved. **Process interdependence** is the interdependence of business processes across organizational boundaries, which denotes that the business processes are passed on to an external specialist who in turn owns, manages, and administers them properly. Companies following this logic may even regulate themselves to two fundamental roles: the position of leader in one set of resources and the secondary role connected to other complementary resources. Still, a strategic goal in these resource coalitions is the choice of orientation and arrangement of an organization in a dynamic resources network. Moreover, Venkatraman and Henderson (1998) underline that resource coalitions demonstrate the very important feature of virtual organizing—fuzzing borders—that is, transcending the borders between competition and cooperation.

The last dimension, **knowledge leverage,** concerns intraorganizational and interorganizational possibilities and means of expertise usage. It is comprised of three stages: harnessing work-unit expertise, recognizing and configuring knowledge as a corporate asset, and gaining access to broad professional community expertise outside an organization. Work-unit expertise covers both designing the technological platform to support unit work as well as organizing and supporting its structures and processes. The second stage concentrates on reaching the expertise in tandem, rather than individually as in previous stage. Here, the cornerstone lies in liberation of tacit knowledge-sharing between individuals within the same organization. In the third stage, the central point of interest is moved up to gaining expertise from the external environment (e.g., other companies, institutions). Organizations endeavor to obtain access to expertise in the extended network and professional networks of external experts.

Scholz's virtual cube model is developed on the expectation that an organization that operates in a turbulent environment is forced to take strategic actions. These activities may be of three types: virtuality realization through the use of ICTs, manufacturing process integration by creating and applying value-added "soft mechanism" (such as vision, shared goals, and trust), and core competence development. According to Scholz (2000), these three different modes and their amalgamation produce unquestionable evidence of the existence of more than one kind of VO. Scholz (2000) postulated that the complex advancement of virtualness could be analyzed in three axes: **core differentiation**, **soft integration**, and **virtual realization.** The configuration of these three theoretical dimensions creates the virtual cube, a representation of virtualness employment (Figure 3.5).

VO members dealing with **core differentiation** have to revolve around their core competencies. In practice, this means that potential VO members must try to develop

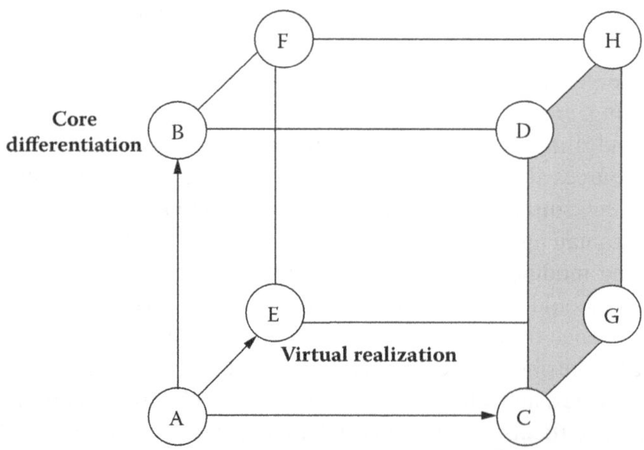

FIGURE 3.5 Virtual cube model. *Source:* Scholz, C. 2000. The virtual corporation: Empirical evidences to a three-dimensional model. *Academy of Management 2000,* Revised Conference Version, Conference Paper, Toronto. With permission.

and provide only those skills and knowledge where they control unique advantages. The discussion of core competencies emanated from the modularity of VO describing the extent to which VO system's constituents can be separated and recombined whenever it is required. Modularity allows the VO system to become flexible by enlarging the number of possible system designs (outputs) attained from a given set of differentiated inputs.

The diversity of core competencies creates various inputs that, in a complex environment with heterogeneous clients, may be discerned as necessary to create a high system capable of capitalizing on a fast-changing market opportunity or demand. The second dimension, **soft integration,** is about executing integration. Integration refers to (1) the degree of tightness of coupling between VO system members and (2) the degree to which system architecture rules enable different combinations of system elements (Schilling 2000). For a VO, it is not only integration that successfully brings together individual parts (VO members) (Scholz 2000). Additionally, two other conditions have to be gratified. First, the independent core competences have to institute a single form that meets a set criterion (e.g., lowest cost, highest quality) established for the whole value chain. Second, the realization of "one face/identity to the client" (a.k.a., "virtual identity") is required. In organizational literature, several suggestions concerning high quality achievement can be found. In the case of VOs, approaches that do not create a new dysfunctional bureaucracy can be used—in particular, vision and shared goals, fairness and trust, and culture of virtuality. Culture of virtuality is understood as a result of bringing together all cultural profiles of VO members. VO chief shared cultural values are both a strict client orientation as well as concentration on technology excellence. VO members usually follow a readiness to work informally, without any formal contract and obligations, for an indeterminate length of time, which in practice is hard to find.

The third dimension, **virtual realization,** serves to create system modularity. It is strongly yoked to ICT and especially with the concepts of virtual reality and cyberspace, which support the internationalization and globalization of business processes.

Finally, it is worth pointing out some very important issues of Scholz's model. Virtualization is perceived similarly as in Venkatraman and Henderson's model (1998) as a strategic option for an organization working in a dynamic environment. It is not coincidence that both differentiation and integration are mostly observed in organizations that operate in constantly changing environments. As a result, a higher degree of virtualization is expected among the companies that are nested in a turbulent environment because these companies' ability to adjust to severity and frequency of changes is key.

REFERENCES

Bultje, R., and Van Wijk, J. 1998. Taxonomy of virtual organisations, based on definitions, characteristics, and typology/ *VoNet: The Newsletter*, 2(3).

Burn, J. M., and Ash, C. 2000. Knowledge management strategies for virtual organisations. *Information Resources Management Journal,* 13(1), 15–23.

Fazlagić, A. 2004. Wiedza jako zasób strategiczny w Banku Światowym. *Bank i Kredyt,* 5, 71–81.

Hale, R., and Whitlam, P. 1997. *Towards the virtual organization*. London: McGraw-Hill.
Keinänen, K., and Oinas-Kukkonen, H. 2001. Virtual organizing as a strategic approach to stay competitive: A conceptual analysis and case study. In *Knowledge Management and Business Model Innovation*, Y. Malhotra (ed.). Hershey, PA: Idea Group Publishing.
Lethbridge, N. 2001. An I-based taxonomy of virtual organisations and the implications for effective management. *Informing Science,* 4(1), 17–24.
McDonough, E., Kahn, K., and Barczak, G. 2001. An investigation of the use of global, virtual and collocated new product development teams. *Journal of Production Innovation Management,* 18, 110-120.
Miles, R. E., and Snow, C. C. 1986. Organisations: New concepts for new forms. *California Management Review,* 28(3), 62–73.
Nilles, J. M. 2003. *Telepraca, strategie kierowania wirtualną załogą*, Warsawa: Wydawnictwo Naukowo-Techniczne.
Palmer, J. W., and Speier, C. H. 1997. *A typology of virtual organisations: An empirical study.* In *Proceedings of the Association for Information Systems 1997 Americas Conference*, Indianapolis, August.
Porter, M. E., and Fuller, M. B. 1986. *Competition in global industries: Coalitions and global strategy.* Boston: Harvard Business School Press.
Prahalad, C. K., and Ramaswamy, V. 2004. *The future of competition: Co-creating unique value with customers.* Boston: Harvard Business School Press.
Schilling, M. A. 2000. Toward a general modular systems theory and its application to interfirm product modularity. *Academy of Management Review,* 25(2), 312–334.
Scholz, C. 2000. The virtual corporation: Empirical evidences to a three-dimensional model. *Academy of Management 2000,* Revised Conference Version, Conference Paper, Toronto.
Skyrme, D. J. 1998. The realities of virtuality. In *Organizational virtualiness: Proceedings of the first VONet workshop*, April 27–28, Bern: Simova Verlag.
Venkatraman, N. 1995. *The IS function in the virtual organization: Who's who wagging whom?* Sixteenth International Confernece on Information Systems, 10-13 December, Amsterdam.
Venkatraman, N., and Henderson, J. C. 1998. Real strategies for virtual organizing. *Sloan Management Review,* 40(1), 33–48.
Wenger, E. 1998. *Communities of practice: Learning, meaning, and Identi*ty. New York: Cambridge University Press.

3.3 VIRTUAL ORGANIZATION MANAGEMENT

VO management pertains to organizing, allocating, and coordinating resources and activities of VOs as well as to interorganizational dependences for purposes of realization within the range of required time, cost, and quality (Karvonen, Salkami, and Ollus 2005, p. 198). VO management varies from management of a conventional organization. It is not that managing a virtual business is entirely different from managing a conventional one (indeed, all the same tasks, such as marketing, financial management, accounting, and operations management, are still de rigueur); however, the nature of these tasks becomes different in VOs. Furthermore, management of VOs is management-plus. It requires adherence to business fundamentals and a grasp of the new environment (Warner and Witzel 2004, p. 1). Therefore, for successful VO management, it is pertinent to know its life cycle. Klüber (1998) presented a wheel depicting a virtual organizing model comprised of four phases: identifying potential partner, a defining strategy, developing appropriate structures, and a coordination mechanism. Analogously, Su, Chen, and Lee (2001) introduced the following four-stage cycle: potential identification, strategy definition, structures development, and coordination mechanism. In addition, these four stages were divided into ten separated steps, which are explained below (Su, Chen, and Lee 2001, p. 67–70).

1. **Marketing analysis and business opportunities capture**
   A company must be able to constantly find existing and forthcoming business opportunities in order to harness them effectively. When a new business opportunity is discovered and cannot be simply dealt with by a single company with required time, cost, and quality criteria, then the VO enters to exploit the opportunity jointly.
2. **Business plan and payoff analysis**
   Before final creation of a VO, the core organization (or information broker) must deliver a business plan and payoff analysis of the potential initiative to give an economic explanation for the existence of the potential VO. In particular, the analyses and calculations should cover the marketing potential, capital capability, technique capability, resource capability, labor capability, and scenarios of possible profitability. Afterward, a report on the business plan and financial analysis should be created to contribute to partner companies' points of reference and decisions about joining or not joining a planned VO.
3. **Competence and demands analysis**
   Once the core company decides to exploit a current business opportunity, the next step must be the competence (satisfiers) and demands (requirements) investigation. First, the capabilities and skills required for this market opportunity must be examined. Second, the capabilities and skills required for this opportunity to be taken must be split up properly into small modules in order to make possible the partner selection process.
4. **Business partner selection**
   On the basis of the previous stage, partner selection is effectuated. Selection criteria must be set in advance and strictly conformed. In addition to capability constraints, other factors related to potential VO members must be also considered, such as operational management level and business reputation.
5. **Cooperation mechanism establishment**
   When the constellation of VO is specified, the cooperation mechanisms among the partners must be established in order to set the seal on the operation efficiency of the whole VO. In particular, the very crucial facets must be considered in this cooperation mechanism, such as responsibility, profit and risk distribution mechanism, quality management, and quality assurance mechanism.
6. **Teams and VO establishment**
   For a VO member, especially a large company, it is not probable to participate as a whole. In this circumstance, a project team (e.g., virtual teams) should be established to participate in the VO and represent the whole company. The VO can be constructed of relatively small organizations or of usually multifunctional project teams to ensure manageable units.

7. **ICT infrastructure establishment**
   In the very beginning of the VO foundation, each partner must construct or advance its current ICT system in order to comply with and back assigned responsibility of its tasks in the VO. Next, all the isolated ICT systems of each VO member must be then attached to an integrated ICT infrastructure for the well functioning of the VO. However, this integrated connectivity makes the system less flexible if one member leaves an alliance.
8. **Virtual operation**
   In the above-described seven stages, several companies and/or teams are brought together to form a single entity called a *virtual organization*. VO operational activities based on ICT infrastructure (but not exclusively) are called *virtual operations*. In contrast to conventional operations, virtual operations exhibit the following three characteristics:
   - The VO enjoys high flexibility. In conventional organizations, a profit is achieved through employing an organization's own resources in an effective manner. A VO meets an end goal through multiple partners' arrangements, integration, and rapid response to a market opportunity or demand. The VO structure and its features constitute the high flexibility of a VO.
   - Both information exchange and interactive forms of communication among VO members are diversified and change rapidly.
   - Control of the entire system is more complex and more difficult than in a traditional company; therefore, it is more fallible. Here comes trust management.
9. **Dynamic review and adjustment**
   In a VO, continuous adjustment to the dynamic changes in the market, the environment, technology, and the organization is imperative. Dynamic review and alteration are vital steps in the effectuation of business goals for which regulations and standards must be established.
10. **VO's disassembly**
    Once the purpose for which a VO was created no longer exists, the deconstruction of the VO becomes the appropriate step.

In terms of VO management, the issue of **instruments ensuring coordination** arises. It is obvious that coordination in traditional organizations is distinct from that of virtual ones (Table 3.4).

Traditional coordination activities grounded in human resources management have given way to coordination activities dealing with knowledge and intellectual management. Generally, there are three types of coordination mechanisms (Reichmann et al. cited in Perechuda 2005): **structural coordination** (empowerment to task specific module), **process coordination** (in the form of plans, rules, programs of activities), and **personal coordination** (exercised by an employee). The fact that the rudimentary construct in gaining integration within a VO is trust cannot be ignored or overlooked.

When debating VO management, it is worth considering the key factors that contribute to the overall success of a VO. As stated by Riemer and Klein (2003) in their model of a successful VO, individual training and learning constitute the human

TABLE 3.4
Reorientation of Traditional Coordination toward Virtual Coordination

Coordination	
Traditional	**Virtual**
Human resources management	Knowledge and human capital management
Command, distribution-based	Competences-based
Acceptance of commands	Consensus building
Superior	Net coordinator (knowledge manager, information broker)
Subjective feature	Process feature
Vertical	Multidirectional
Task coordination	Processes coordination
Formal authority	Authority based on core competences and reliability
Control	Trust
Hierarchy	Tasks

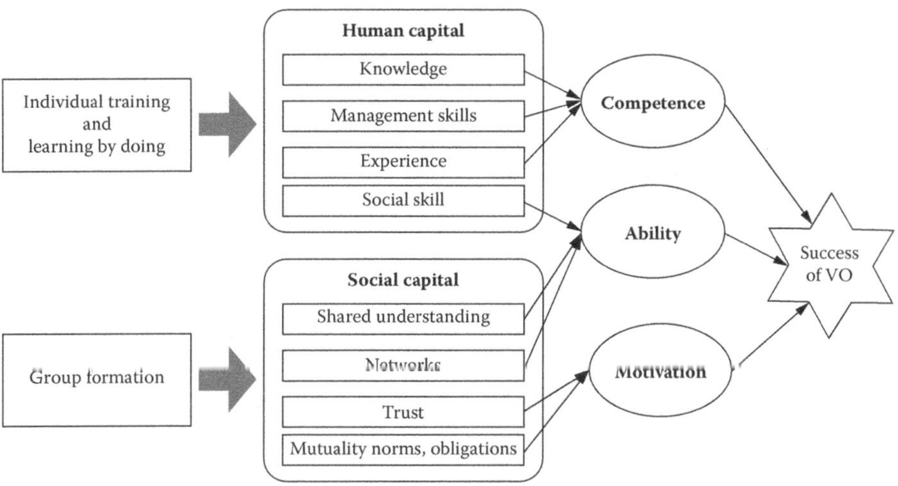

FIGURE 3.6 Success factors for VOs based on Riemer and Klein (2003).

capital, whereas group formation is regarded as social capital. These two major constructs enhance each other and are embodied in competence, ability, and motivation in the organization, therefore conditioning successful performance of a VO. The role of ICTs in developing human capital is not significant. At the same time, social capital cannot be overvalued (Figure 3.6).

Success for a VO may be supported by putting into practice **VO management methods** that a VO manager or VO member uses in order to achieve the VO's objectives within certain time, cost, and quality frames. (Negretto 2005, p. 17). There are five approaches to VO management (Table 3.5). Each approach is framed by application area, management focus, and techniques (Negretto 2005, p. 21–22).

TABLE 3.5
VO Management Approaches

Approach	Application Area	Main Focus	Techniques to be Applied
Multi-organizational project management	A specific task, a defined goal. Constraints with outcome cost and time. High/medium risk, considerable interdependencies of tasks/operations (capital projects), possibly strong technical dependencies.	Achieving an integrated result or objective. Keeping the schedule and costs. Efficiency is important.	Project management techniques, e.g., critical path analysis etc. Effective/medium operational/technical coordination: predefined practices and processes for approving and checking intermediate results, defined technical interface. Planning of activities, allocation of resources, allocation of responsibilities, and distribution of management. Following a plan, progress reports, auditing, monitoring and real-time performance measurement. Monitoring of results/activities/inputs. Configurable/tailored plans/plan templates of one-of-a-kind VOs.
Encouragement approach	Abstract objective/changing objective/not bounded objective, first stages of innovation process, research, business development.	Quality of end results, innovation, often within time and cost constraints. Management of scope and motivation.	Clarification of purpose, building of common vision, direction, discussion, conclusion, consensus building. Monitoring/measuring motivation, engagement and performance. Empowerment and feedback. Responding to human needs: desire for knowledge and learning, significance, belonging to a group, acceptance, self-realization.

Self-organizing approach	All the partners have internalized their objectives—the objectives of partners support each other, no potential conflicts and unnecessary dependencies.	Management of motivation and results, integration may be applicable.	No central coordination. Rules for profit, profit- and knowledge-sharing are needed.
Time-dominated VOs	Fast, emerging VOs with a short life cycle, time of set-up often not known in advance. Emergency and maintenance systems. (Semi)automated service tasks. Partners have a pre-agreement of the objective.	Being able to complete task as fast as needed. (Being fast means prevention from potential losses.) Costs are not as important as in other cases.	Standard detailed pre-plans. No time for planning or fast problem identification phase. Coordination and responsibilities defined in advance. In some cases depending on task: automated/semi-automated monitoring and execution, (semi) automatic negotiations and decision making, predefined solutions, transferring the decision power to the automated system.
Supply chain management approach	Supply chains, usually longer duration (i.e., product or product life cycle)	Orders and deliveries, logistics, efficiency and time.	Forecasting, allocation of resources, planning, monitoring, reallocation, etc.

Source: Negretto, U. 2005. *Report on methodologies, processes, and services for VO Management*. ECOLEAD

The most well-known traditional management approach which can be applied to VOs is the **multiorganizational project management approach.** This approach usually requires well-defined goals as well as progress monitoring. Another approach, **supply chain management,** is applicable when a VO is originated to deliver certain products or services in a limited period of time.

The **encouragement approach**, as distinct from the multiorganizational project management approach, and does not possess such clear, tangible, and stabile goals. On the one hand, typically there are constraints for time and costs but, on the other hand, the end result may have no upper limit, and it may even be nonviable to know in advance what a VO brings off. Sometimes, the highest nonforeseen success is achieved when utterly new products and services are created. In this type of VO, it is important to emphasize the importance of following time and cost restrictions; therefore, some project management techniques are relevant. But to be successful, the main approach in VO management is to ensure that all partners work toward attaining the shared purpose in their scope of responsibilities. Just as it is important to motivate, it is equally important to ensure that each member is working toward fulfilling the shared purpose.

The **self-organizing approach** is appropriate when no coordination is required. Each partner performs its own initiatives in collaboration toward the common goal. In this situation, each VO participant has clearly internalized the VO purpose, there are no minor conflicts, and there are commonly accepted rules of collaboration. Usually, it implies that none of the partners accept another partner's risk and each is responsible individually for caused losses. In this approach, project management might be used to support monitoring and profit-sharing. When there is no appointed VO manager, the VO partners must establish a common deal and rules concerning performance management. What is more, in most cases, VO members are dependent on the performance of others. Thus, the self-organizing approach is neither fitting nor compatible with circumstances in which high losses are likely if the VO's goal is not achieved.

The **time-dominated VO** is a specific example of time requirements that, in turn, lead to the specific application of management methods. The need for a time-dominated VO emanates from two scenarios: emergency management and (semi) automatic service requests. In these fast situations, it might be difficult to quickly form and activate a VO. The same could be said about the implementation of real-performance measurement to support the VO management. In this case, the role of performance measurement is even more important than after the dissolution phase, when it mostly aims at experience gathering and learning from the VO are the most important goals.

Moving on, it is worth pointing out here that **planning** in VO management does not differ significantly from that in conventional project management. As a result, one is allowed to apply some techniques acquired from project management.* Examples are: GATT charts, CPM (critical path method), PERT (program evaluation and review technique), financial planning, budgeting, and profitability analysis. To grasp

* *Project management* is "those processes of managing that change by planning the work, executing it, and co-ordinating the contribution of the people and organizations with an interest in the project" (Slack 1998, p. 156).

a VO's reality, one has to focus actions on risk management, in particular, financial risk management, feasibility risk, and competence risk.

In a VO, continuous monitoring of activities is necessary to get a reliable picture about the present state of task accomplishment in terms of time, quality, and costs, the chance of VO in order to evaluate the chance of a shared goal achievement. Along with that comes the control of time, quality, and costs. While this is very difficult in a dynamic environment, the following techniques can be used:

- Real-time performance measurement connected with data acquisition, normalization, calculation, trend analysis, projection, and visualization; measures might include used resources, percentage of completed deliveries, and earned value
- Simulations
- Discussions and interviews with VO partners (face to face, phone, Web meetings)
- Visits to VO partners' facilities
- Visual observations of progress
- Auditing and control of reaching milestones
- Online monitors and Web cameras at VO partners' sites and delivery sites. (Negretto 2005, p.18)

On the basis of observation activities, a VO manager should expect probable deviation from the initial plan and have ready relevant corrective actions. There are many options. Table 3.6 contains examples of tools that might be used for corrective actions. This is only a brief outline of the topic. For a more comprehensive description, interested readers are referred to the rudimentary literature on management.

Given the importance in this section to the approaches used in VO management, it is worth citing empirical studies that show what management ideas are used by VOs. Here we can invoke the findings of surveys concerning the management concepts used by Polish companies, which put the interorganizational VO idea into practice according to Definition 3.1. (Graph 3.3).

A large majority of companies (70%) that apply VO forms in business practice employ project management as well. This finding corresponds with the literature on the VO management approach discussed in this section and highlights the importance of project management for VO management. It is striking evidence that it is important for VOs to find a framework that enables them to monitor the pursuit of the common goal. Other concepts revealed by the survey to support VOs are management by objectives, team building, and knowledge management—which can be observed respectively in 55%, 52%, and 48% of the surveyed companies.

Moreover, it should be emphasized that effective functioning of a VO requires a specific arrangement among many interrelated organizations, which strongly depends on the ability to communicate within the definite constellation (Cygler 2002, cited in Najda 2003, p. 81). Therefore, team building is reported in 52% of surveyed companies applying the VO idea.

TABLE 3.6
Examples of VO Management Methods for Corrective Actions

Management Methods	Type of Issues/ Plan Deviation	VO Consideration
Force power, authority	Time, quality	Difficult to implement. Can be used in conjunction with economic threats or punishment.
Motivation	Any	Not a stand-alone method. To be used with or as a result of other actions. Difficult to implement in general, even more difficult in VOs.
Punishment	Any	Provisions for punishment must be agreed upon in the VBE or in the VO contract.
Rewarding	Costs, time	Usually financial.
Replanning schedule, resources allocation and responsibilities	Costs, time	Difficult without force/power. The VO contract limits. Multi-VO environment limits the applicability.
Review and modification of VO's objectives	Quality	Can be used in VO with unclear objectives (e.g.. product development VO).
Training, guidance, templates	Quality	Difficult to implement in short-lived VOs. Time and cost aspects may limit the usage.
Communication, team building	Time	More important than in single-company environment, where this is part of company's development activities.
Creativity management	Quality, costs	Pivotal for growth of VO's competitiveness.
Strategic insensitive	Any	Long-term strategic insensitive for future collaboration is a driver and motivation for good performance.
Image and reputation building	Any	Close to honoring and rewarding. Suitable for long-term VOs.

Source: Negretto, U. 2005. *Report on methodologies, processes, and services for VO Management.* ECOLEAD, p. 30.

The main challenge for VO management issues from the temporary nature of the VO and distribution of its functions among organizations. Of all the traits traced in VOs, one is strategic: the VO must be very responsive to changes in an environment. Thus, it is necessary to apply a dynamic approach to VO management, such as that described by Mowshowitz (2002). It is perfectly possible by the restructuring of hitherto existing conventional approaches toward management. On top of that, it is evident that managing VOs involves real-time performance measurement and therefore overall performance management.

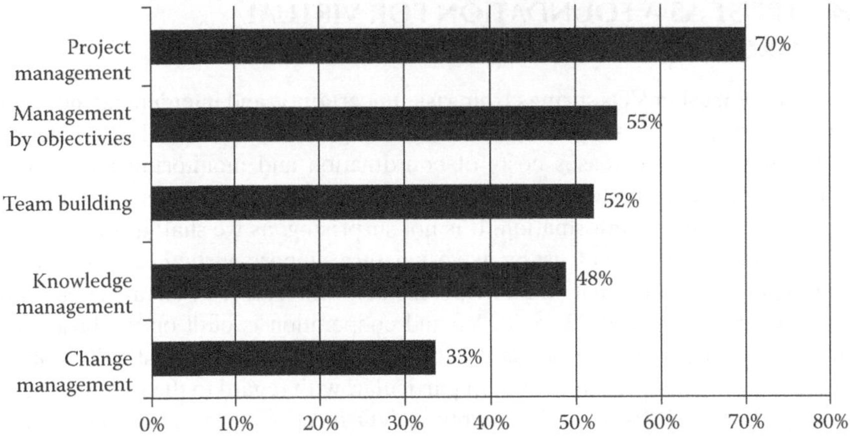

GRAPH 3.3 The employment of management concepts in companies applying the inter-organizational VO idea (n = 33). Please note that answers do not add up to 100% because respondents were able to give multiple answers to this question. *Source:* Sankowska, A., and Wańtuchowicz, M. 2006. *Organizacja wirtualna jako forma organizacji sprzyjająca rozwojowi przedsiębiorstwa.* Unpublished Statutory Works, Faculty of Production Engineering, Warsaw University of Technology, Warsaw. With permission.

REFERENCES

Karvonen, I., Salkari, I., and Ollus, M. 2005. Characterizing virtual organizations and their management. In *Collaborative networks and their breeding environments,* L. M. Camarinha-Matos, H. Afsarmanesh, and A. Ortiz (eds.). New York: Springer.

Klüber, R. 1998. A framework for virtual organizing, In *Organizational virtualness: Proceedings of the VoNet workshop,* P. Sieber and J. Griese (eds.), April 27–28. Bern: Simowa Verlag.

Mowshowitz, A. 2002. *Virtual organization: Toward a theory of societal transformation stimulated by information technology.* Westport, CT: Quorum Books.

Najda, M. 2003. Zarządzanie organizacją wirtualną. In *Zarządzanie wartością firmy w dobie kryzysu,* S. Kasiewicz and L. Pawłowicz (eds.), 79–83. Warsaw: CeDeWu.

Negretto, U. 2005. *Report on methodologies, processes, and services for VO Management.* ECOLEAD.

Perechuda, K. 2005. Instrumenty koordynacji w przedsiębiorstwie wirtualnym. In *Komputer: Przyjaciel czy wróg?* A. Szewczyk (ed.), 401–406. Uniwersytet Szczeciński, Międzyzdroje-Szczecin.

Riemer, K., and Klein, S. 2003. Challenges of ICT-enabled virtual organisations: A social capital perspective. In *Proceedings of the 14th Australasian Conference of Information Systems (ACIS),* Perth.

Sankowska, A., and Wańtuchowicz, M. 2006. *Organizacja wirtualna jako forma organizacji sprzyjająca rozwojowi przedsiębiorstwa.* Unpublished Statutory Works, Faculty of Production Engineering, Warsaw University of Technology, Warsaw.

Slack, N. 1998. *The Blackwell Encyclopedic Dictionary of Operations Management.* Oxford: Blackwell Business.

Su, Q., Chen, J., and Lee, S.-M.. 2001. Quality management system's design for virtual organizations. *Electronic Journal of Organizational Virtualness,* 3(5), 65–78.

Warner, M., and Witzel, M. 2004. *Managing in virtual organizations.* London: Thomson.

3.4 TRUST AS A FOUNDATION FOR VIRTUAL ORGANIZATION FUNCTIONING

The need for trust in VOs springs from risk, uncertainty, and interdependence among partners. Trust can, if employed correctly, bring benefits concerning efficiency of cooperation. Trust decreases costs of coordination and monitoring and enhances creativity. Partners who trust each other are able to collaborate, even conditioned by the limited amount of information. It is not surprising, as we shall argue repeatedly throughout this book, that trust by its very nature supports virtual cooperation. This in turn leads to the idea that "trust [...] is the defining feature of virtual cooperation" (Davidow and Malone 1992, p. 9). Virtual cooperation is built on the basis of key competencies, though what bonds it is the trust itself. Trust dramatically improves the adaptability of an organization, in particular, with regard to the pace of response to change (Bavec 2004, p. 403). Doubtless, VO success relies substantially on trust building and sustaining. Trust is the focal issue due to the fact that it sets out features of virtual cooperation. When valuing trust in a VO, it is inevitable to point out that trust in a VO acts as a substitute for both hierarchical control and the contractual agreement required for the regulation of its creation, operation, and termination. Many authors underscore the role of trust in relations based on cooperation (e.g., Casson 1991, Dasgupta 1988, Gambetta 1988, Handy 1995, Lorenz 1988, Ring and Van de Ven 1992). Researchers perceive trust as the most important component contributing to the success of collaboration, strategic alliances, and partnering networks of small- and medium-sized companies.

A survey conducted by Grudzewski, Sankowska, and Wańtuchowicz (2005) among 11 experts from small- and medium-sized Polish companies from the IT sector experienced in ad hoc cooperation proved the aforementioned conjecture widely spread in the literature. From 48 initial factors that are believed to impact the decision to select or not select a partner for a potential VO, trust received the highest sum of ranks given by experts. Each factor could receive 44 points maximally. Trust was ranked highest among 10 critical success factors* for VOs with the highest sum of ranks (Graph 3.4).

With Graph 3.4, it becomes apparent that trust is the critical factor for VO success. Reliance on trust in VOs between "strangers" or "relative strangers" that are part of a network is essential to an even greater extent than in traditional organizations. This justifies the assumption that trust is the factor of greatest importance for VOs. Trust is, in the figurative sense, the heartbeat of a VO.

On the grounds of similar research carried out in the United States (Grudzewski, Sankowska, and Wańtuchowicz, in press) among experts experienced in intercollaboration (Graph 3.5), we can acknowledge trust as the most critical factor for VO success, too. The findings of this research imply that the eminent importance of trust in VOs is easily recognized both by Polish and American experts.

In an analogical expert survey carried out in Sweden (Graph 3.6), the highest ranks were given to honesty and clearness of communication, which as a matter of fact is a precondition of a trust atmosphere between collaborating parts. At very first

* Critical success factors are "the limited number of areas in which results, if they are satisfactory, will ensure successful competitive performance for the organization" (Rockart 1979, p. 85).

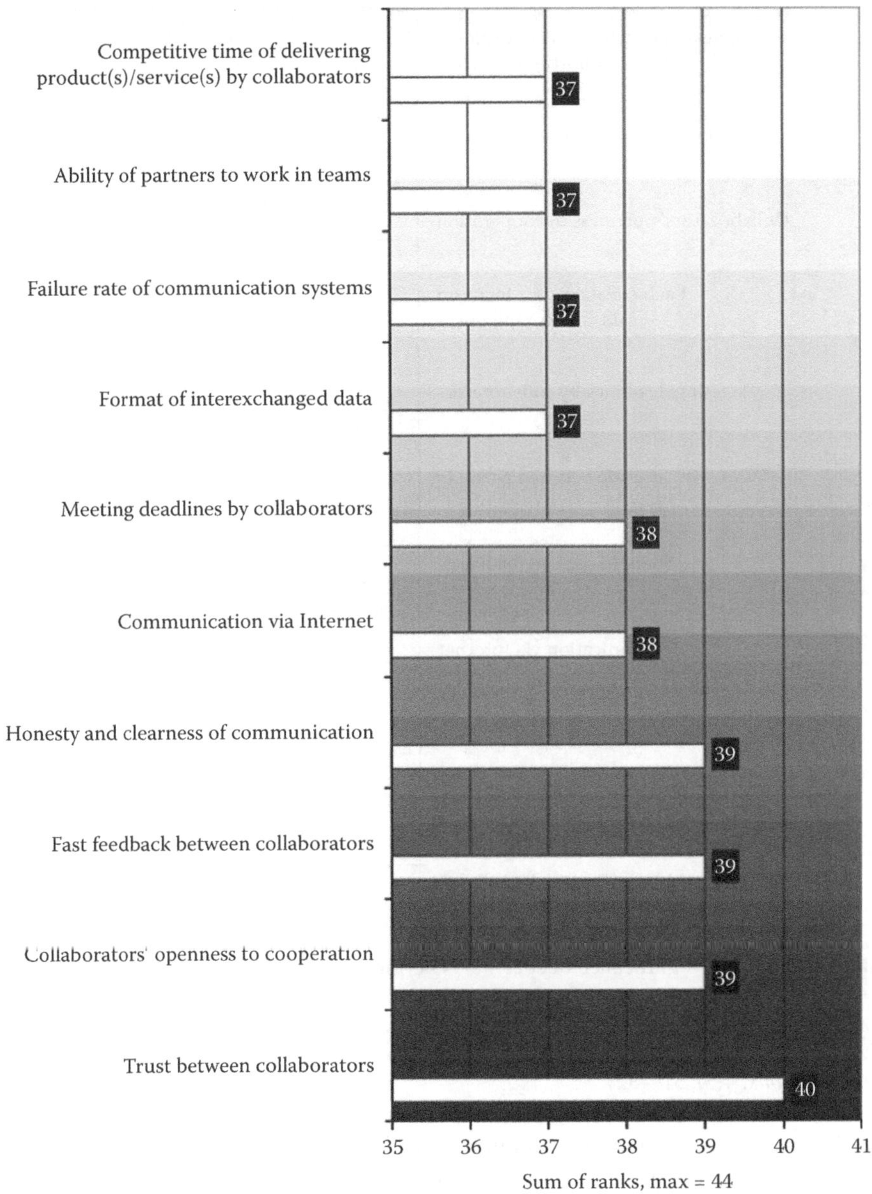

GRAPH 3.4 The significance of specific VO critical success factors by Polish experts. *Source:* Based on Grudzewski, W. M., Sankowska, A., and Wańtuchowicz, M. 2005. Virtual scorecard as a decision-making tool in creating virtual organisation. In *Collaborative networks and their breeding environments,* L. M. Camarinha-Matos, H. Afsarmanesh, and A. Ortiz (eds.). New York: Springer, p. 257.

glance, it can be surprising that trust took fifth place in the Swedish ranking. This can be accounted for by the fact that Swedish culture is among the most trusting cultures in the world. The Swedish face high trust every day, so they naturally devote

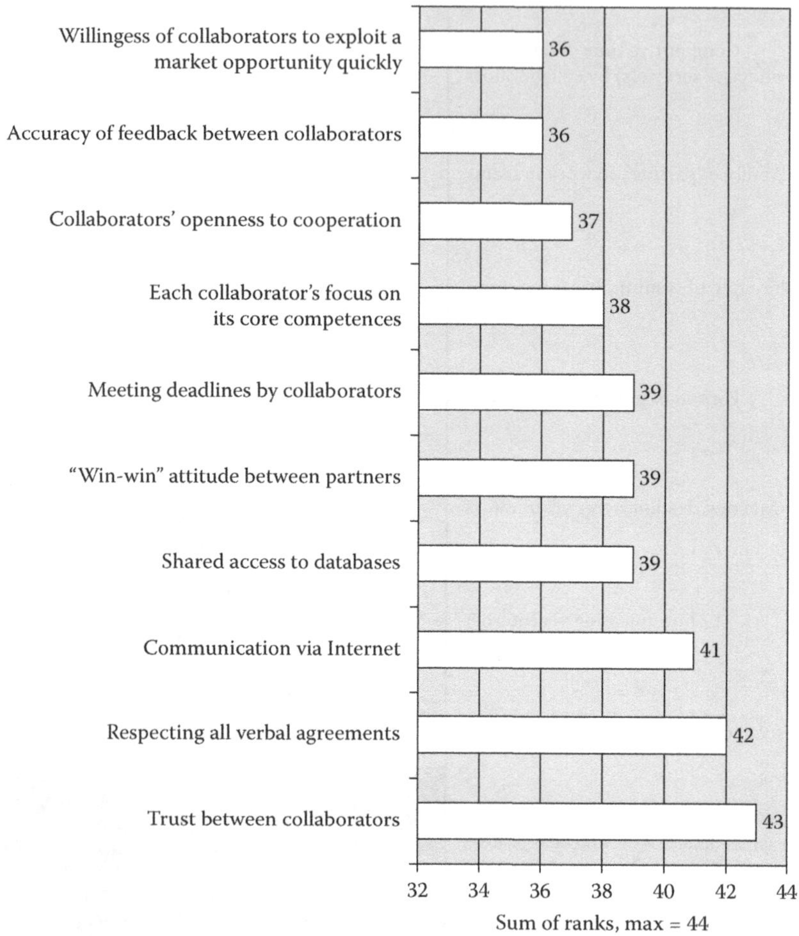

GRAPH 3.5 The significance of specific VO critical success factors by American experts. *Source:* Based on Grudzewski, W. M., Sankowska, A., and Wańtuchowicz, M. (in press). A comparative analysis of experts' perception of critical success factors affecting virtual organization: North American versus Polish. *International Journal of Networking and Virtual Organisations*, 4(3), 318–327.

much care to maintaining its prerequisites, which are honesty and clarity of communication. Moreover, the Swedish ranking sheds light on other issues. Seven out of ten critical success factors for VOs in the ranking are connected with the preconditions of trust and communication: honesty and clarity of communication, communication via the Internet, security of ICT systems, ability to communicate in foreign languages, high throughput and transfer rate of communication systems, communication via fixed phone, and communication via mobile phone. Given this data, we can state that trust is a critical success factor for VOs in the Swedish ranking, too. We can even advance the thesis that a successful VO results from two decisive phenomena— namely, process approach (which was discussed in the section entitled "The Multifaceted Nature of Virtual Organizations") and high level of trust (Figure 3.7).

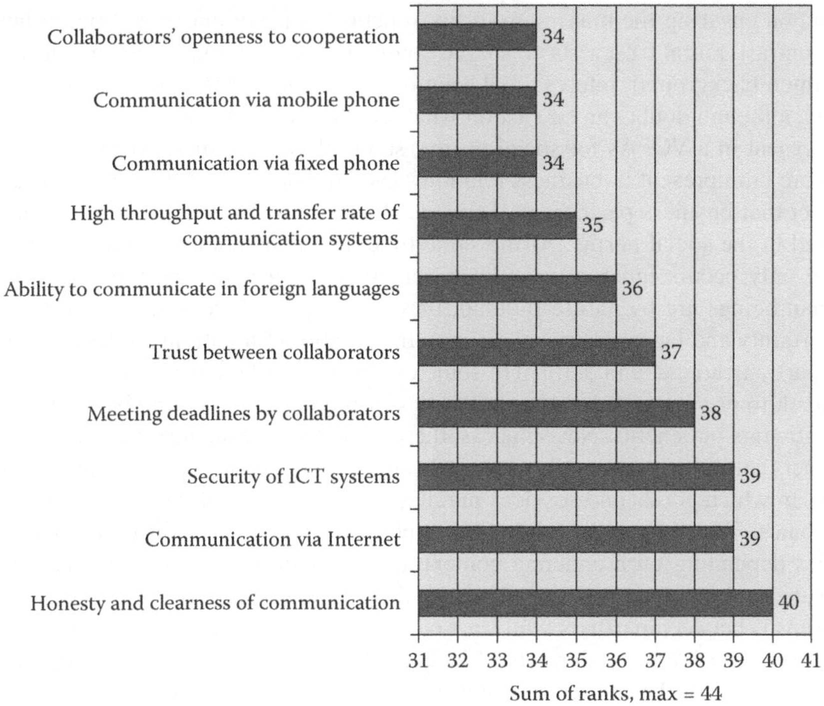

GRAPH 3.6 The significance of specific VO critical success factors by Swedish experts.

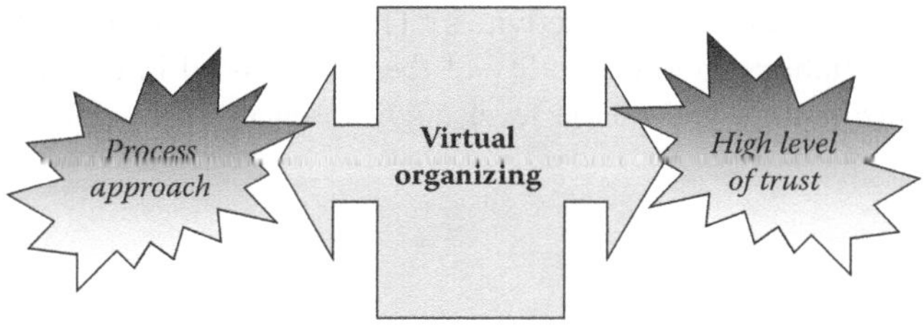

FIGURE 3.7 Critical elements of virtual organizing.

Only trust can instill subjective confidence in the partners' good intentions and their abilities to fulfill obligations. At the same time, experts have expressed concern about obstacles in trust establishment (Grudzewski et al. 2005). Thus, the literature and empirical insight into trust is thoroughly substantiated. It may lead to the emergence of VOs in organizational practice.

As was noted in the section about the notion of trust, we can fundamentally distinguish two main trust streams: **rational perspective,** centered on the calculus of self-interest, and **social perspective,** which revolves around moral duty (Ishaya and Macaulay 1999; Jarvenpaa, Knoll, and Leidner 1998; Uslaner 2003) and shared values. The former perspective perceives man as a rational actor who chooses a course of action ensuring the maximum utility function. It is similar to betting on horses. By contrast, moral trust rests on shared common values, norms, kinship, familiarity, common background, interest, and even religion. Both types of trust might exist in a VO, although doubt can be cast on whether trust derived from morality is in reality present in a VO. As for social norms, surely they coordinate social interactions that are omnipresent in business relations, even remote ones. Albeit, it is difficult to expect that business partners will trust each other just because this requirement is rooted in the social norms. At the same time, it is contentious to state that in VOs there only occurs inherent calculus-based trust. Even assuming theoretically that human beings are by nature rational, they must make choices and act in terms of uncertainty and face resource (e.g., time, information) limitations, so their decisions are partly irrational and defined to some extent by their beliefs about the world. It is difficult to predict each VO member behavior relying solely on the basis of alternative streams of benefits. Nevertheless, there is empirical evidence that trust between partners can appear even without shared values. An example of that might be e-shopping, in which goods and services purchasing is made in order to achieve expected outcomes. Trust in a VO is therefore rather a combination of different motivating forces depending on a concrete context and cooperating partners' characteristics. To sum up this issue, trust is built on continuous, ongoing step-by-step incremental consensus between partners who start out with very imperfect knowledge and inexperience of each other (Ishaya and Macaulay 1999, p. 144–145) and depends heavily on the context of the cooperation (e.g., spatial, temporal, cultural background).

By the same token, Ishaya and Macaulay (1999) identified three levels of trust in virtual teams (which was noted as a form of organizing work within a VO): **technology level, media level,** and **social level**, which are presented in Figure 3.8.

The first two levels of trust, technology and media, are the technique and "software" facilitating collaboration activities. The media level of trust rests on the functionality, reliability, and capability of the system used in communication processes. The technology and media level ought to ensure the security of conveyed

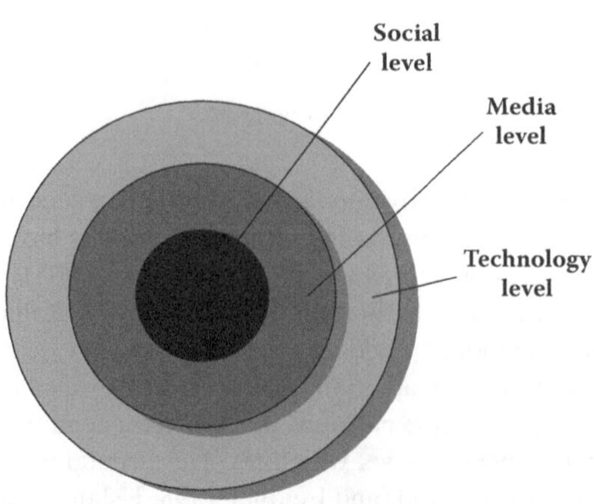

FIGURE 3.8 Difficulty of achieving certain types of trust.

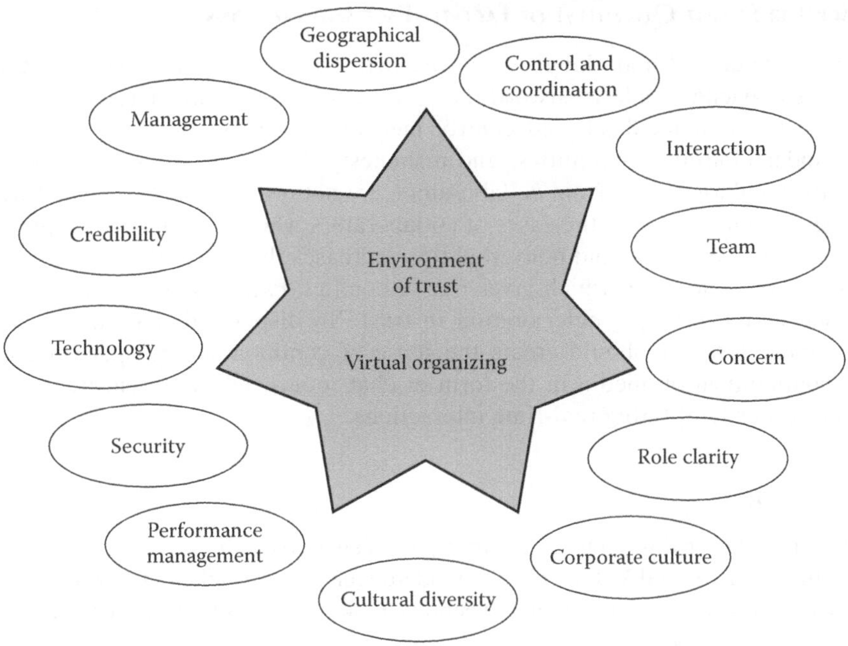

FIGURE 3.9 Virtual circle of trust. *Source:* Gujar, A., and Malm, T. 2005. *A conceptual framework of trust in virtual teams*, 26th Annual Industrial/Organizational–Organizational Behavior Conference, Melbourne, FL. http://www.fit.edu/psych/3t/papers.php (accessed November 3, 2006), p. 9. With permission.

information (i.e., accessibility, confidentiality, access control, authentication, detectability of "information leakage"). This type of trust is to a large extent important to be ensured in nonadvanced ICT environments.

Media and technology levels are less significant for trust than the social level. Social trust is more intricate and, for this reason, more difficult to build. Nevertheless, one can point out some things that create or destroy social trust. Among factors which undercut trust between parties might be the following: flaming, making unreasonable requests from people, ignoring requests, badmouthing, and failing to meet obligations and deadlines. Those things usually come into sight in daily, face-to-face interactions between people, although there are many things that one perhaps would not say to another; whereas, in virtual communication, people might easily go on criticizing others for long periods. They are also more likely to say things they would not have the courage to reveal in face-to-face communication (Ishaya and Macaulay 1999). Another important aspect is to notice that trust tends to deepen or retrogress as a consequence of mutual expectations and experience and, in turn, this influences outcomes.

Having discussed the magnitude of trust, it is high time to discuss major factors that build and retain trust and present a challenge for processes concerning trust creation in terms of virtual cooperation. These factors, according to Gujar and Malm (2005), are described below (see also Figure 3.9).

Lack (or Slight Quantity) of Face-to-Face Interactions

Human voice and body language ensure permanent feedback about a speaker's feelings, background, trustworthiness, capability, and understanding. Without such social feedback and control mechanisms, it is complicated to avoid misunderstanding, ambiguities, and dishonesty. Social needs, such as psychic comfort and assurance from collaborators, are not often present in virtual working environments, as in the cases of collaborators who work in physical proximity. The lack of verbal and nonverbal tips decreases the probability of successful transfer of a message, which gives rise to conflict. Failure in conflict resolution in succession prompts deterioration in trust. To disperse damaging effects of proximity loss, we should create the sense of common shared space rendered by communication media in the form of chat rooms, teleconferences, or video conferences facilitating real-time interactions.

Technology

Problems concerning technology, to repeat, can impact the effectiveness of geographically dispersed VO members. Standard technology might be far from satisfactory in terms of use or economic feasibility required for VO. Moreover, training in use of these technologies may generate additional complications.

Geographical Dispersion

This issue refers to virtual workers perceived as being "out of sight, out of mind" from the standpoint of their organizations as well as their attitude toward their organizations. This is a potential hazard in trust development since VO members can retreat from each other in terms of engagement, identification, and trust as they experience lack of direct social contact.

Security

VO members might reluctantly transfer information as a consequence of distrust in security systems (in particular, their confidentiality). Harrington and Rumpel (1999) (cited in Gujar and Malm 2005) proposed employment of formal means of control or security measures, such as passwords, electronic encryption, or electronic signatures to protect proprietary information.

Shift in Control and Coordination

Since VOs have adapted forms of networking and horizontal communication between members, traditional (hierarchical) methods of coordinating activities seem to be not applicable. Direct supervision and enforcement of rules are not relevant in the context of VOs. On that account, trust and interdependence are substitutes for formal methods of control in VOs.

Performance Measure

VOs manager must often appraise outcomes and skills from a distance. Conventional control methods are rooted in direct control and scheduling and require on site access to workers, which is difficult to accomplish at a distance. Moreover, for VOs, hierarchical structure is a pain in the neck. As a result, autonomy, networks based on trust, integrity, and well-thought-out security plans are necessary to ensure the realization of tasks.

Team Existence

VOs and teams must be permanently redesigned in order to harness a short-run market occasion. As soon as a certain business goal is accomplished, a VO is dismantled. In such cases where there are not embodied long-term commitment or loyalty, cynicism can develop.

Role Problems

Fundamentally in VOs, weight is given to differences in competencies rather than role specifications. Therefore, it appears that inconsistency in role behavior and "blurring" of roles can result in trust erosion (Jarvenpaa and Leidner 1999, as cited in Gujar and Malm 2005). Additionally, VO members might come under different authorities, which decreases the transparency of activities.

Cultural Diversity

The dimensions of individualism versus collectivism and power versus distance shape trust appreciably. The other dimensions of culture pertain to uncertainty avoidance and high context versus low context, and they mold VO members attitudes toward their needs, values, and goals. If VO members are not in accord concerning values, need, and goals, serious problems can arise.

Corporate Culture

The stimulus for cooperation, norms of equity, orientation on people, flexibility, and participation of rank-and-file in decision making, high quality of feedback processes, communication, support and concern for others are fundamental elements enhancing trust.

Management

Rigid management attitudes could be a barrier to VO creation. Once a manager accepts that he or she cannot manage something that he or she cannot directly observe and then if an employee "out of sight" acts opportunistically, trust between virtual team members and the leader is undermined. On the contrary, frequent, accurate, and open communication between supervisors and subordinates builds trust.

Credibility

Approval of VO high members' professionalism created competencies-based trust. By contrast, inexperience as well as the unreliability of partners constitutes a lack of trust.

Integrity

Perceived integrity, when adhered to the perception of performance, builds an overall picture of parties. The temporal nature of a VO makes integrity a stipulation in the processes of trust formation.

Concern

Concern over the welfare of a partner is a landmark in trust building between parties. Once again, trust is a very critical success factor for VOs. From the VO facet, the sizeable literature is in agreement here. However, trust is very complex and an "umbrella" construct begging to be given greater importance than it has been given so far by management sciences. Especially for VOs, trust is really hard to establish because due to their temporal nature, VOs require a high level of trust establishment, often in a very short-time life cycle. In this light, the conceptualization of trust from a management perspective is needed for VOs. It is therefore not surprising that trust management is a vital issue and enabler for VOs (Grudzewski, Sankowska, and Wańtuchowicz 2006). Trust management is a bespoken challenge to management science that should be studied and promoted nowadays. The matter of trust in VOs shall be discussed in more detail in the section entitled "Case Studies of Virtual Organizations."

REFERENCES

Bavec, C. 2004. On management in virtual environments: Between trust and control. In *Intellectual capital and knowledge management,* Proceedings of the Fifth International Conference of the Faculty of Management Koper, University of Primorska, Slovenia, 397–403. http://www.fm-kp.si/zalozba/ISBN/961-6486-71-3/397-404.pdf (accessed November 3, 2006).

Casson, M. 1991. *The economics of business culture: Game theory, transaction costs, and economic performance.* Oxford: Claredon.

Dasgupta, P. 1988. Trust as commodity. In *Trust: Making and breaking cooperative relations,* D. Gambetta (ed.). New York: Blackwell.

Davidow, W. H., and Malone, M. S. 1992. *The virtual corporation.* New York: Burlingame Books.

Gambetta, D. 1988. *Trust: Making and breaking cooperative relations.* Oxford: Blackwell.

Grudzewski, W. M., Sankowska, A., and Wańtuchowicz, M. 2005. Virtual scorecard as a decision-making tool in creating virtual organisation. In *Collaborative networks and their breeding environments,* L. M. Camarinha-Matos, H. Afsarmanesh, and A. Ortiz (eds.), 293–300. New York: Springer.

Grudzewski, W. M., Sankowska, A., and Wańtuchowicz, M. 2006. Trust management in virtual organisations of small- and medium-sized enterprises, *Economic and Organization of Enterprise,* 1(672), 3–9.

Grudzewski, W. M., Sankowska, A., and Wańtuchowicz, M. (2007). A comparative analysis of experts' perception of critical success factors affecting virtual organization: North American versus Polish. *International Journal of Networking and Virtual Organisations*, 4(3), 318–327.

Gujar, A., and Malm, T. 2005. *A conceptual framework of trust in virtual teams*, 26th Annual Industrial/Organizational–Organizational Behavior Conference, Melbourne, FL. http://www.fit.edu/psych/3t/papers.php (accessed November 3, 2006).

Handy, C. 1995. Trust and virtual organization. *Harvard Business Review,* May-June, 40–50.

Ishaya, T., and Macaulay, L. 1999. The role of trust in virtual teams. *Organizational Virtualness and Electronic Commerce*, 1(1), 140–157.

Jarvenpaa, S. L., Knoll, K., and Leidner, D. 1998. Is anybody out there? Antecedents of trust in global virtual teams. *Journal of Management Information Systems*, 14(4), 29–64.

Lorenz, E. H. 1988. Neither friends nor strangers: Informal networks of subcontracting in French industry. In *Trust: Making and breaking cooperative relations,* D. Gambetta (ed.). New York: Blackwell.

Ring, P. S., and Van de Ven, A. H. 1992. Structuring cooperative relationships between organizations, *Strategic Management Journal*, 13(7), 483–498.

Rockart, J. R. 1979. Chief executives define their own data needs. *Harvard Business Review,* March–April, 57(2), 81–93.

Uslaner, E. M. 2003. Trust in the knowledge society. Conference on Social Capital, Cabinet of Government of Japan.

4 Practical Strategies for Trust Management in Virtual Organizations

4.1 TRUST AS THE PRINCIPAL COMPONENT OF ORGANIZATIONAL STRATEGY

Trust in Organizations

Trust is always associated with building and functioning of an organization. For example, in organization and management theory, Henri Fayol (1949) pointed out that crucial issues include good atmosphere among colleagues, *esprit de corps,* and trust. Trust played the foremost role in the past, but it is still a meaningful phenomenon in contemporary organizations. Interest in trust grows mutually with a higher and higher number of strategic alliances and other forms of collaboration. It can be discerned that lack of trust results in an increase of 50% in the costs of business activity (Ryan 1996). Mistrust toward people evolves into the need for control and monitoring of their behaviors, which create additional costs. Doing business in tumultuous times indissolubly leads to trust, which becomes an essential factor in strategy and condition for existence of an organization.

In organization and management science, trust is predominantly explained in multidimensional conceptions. McAllister (1995) (see the Chapter 1 subsections "Theoretical Approaches toward Trust" and "Methodology of Trust Measurement") defines **organizational trust** as the extent to which a person is willing to act on the basis of, the words, actions, and decisions of another. He identifies two types of organizational trust: cognition-based (rational) and affect-based (emotional). The former originates from an individual's beliefs (care about peers, commitment). The latter is determined by the rational and reasonable circumstances.

In contrast, Galford and Drapeau (2003) crystallized the concept of **organizational trust** as the trust people have not in any individual (supervisor, colleague, leader, manager) but in the company itself. The many issues associated with organizational trust include:

- Does the company keep promises?
- Are processes fair and consistent?
- Are organizational goals consistent with goals of groups within the organization?

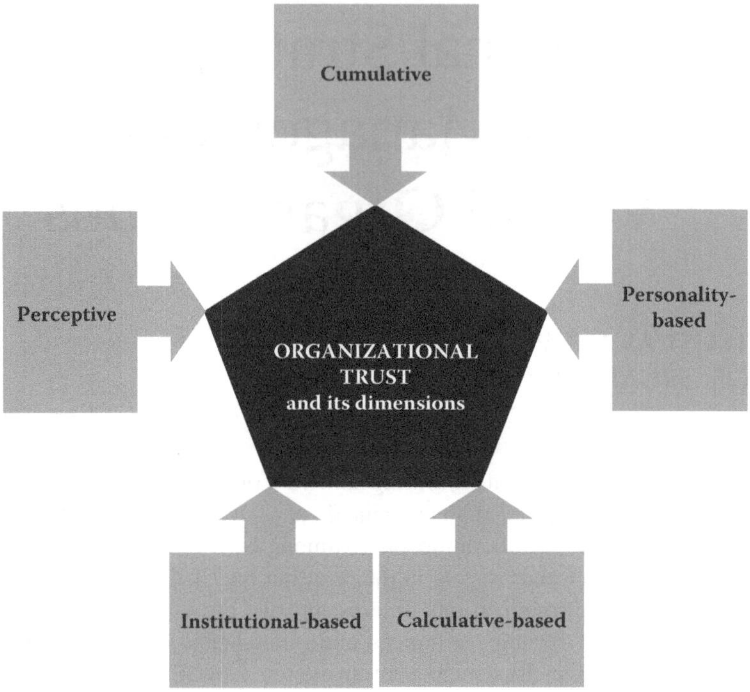

FIGURE 4.1 Organizational trust dimensions. *Source:* Kordel, P. 2002. Zaufanie a proces tworzenia strategii przedsiębiorstwa, *Organizacja i kierowanie*, 2(108), p. 79.

Maintaining organizational trust is complex and depends on many various circumstances.

On the grounds of management literature, we can extract five major dimensions of **organizational trust** (Figure 4.1) in the trust-measurement and trust-building processes.

The **personality-based dimension of trust** in organizations begets the system in which norms and values are shared by all members of the organization. Furthermore, focal place have propensity to trust and personality traits of certain individual that are pertinent to this dimension. The **calculative-based dimension of trust** comes into play when an individual compares profits and losses generated during a trusting relationship (Kordel 2002). During the trust-building process, there is first and foremost a need for fairness and a stable environment—in other words, **institutional-based trust** (Bratnicki and Kordel 2000). A mutual trust relationship and its perception by business partners are included in the **perceptive dimension of trust**. The last dimension of trust, **cumulative,** grows out of the knowledge possessed by relationship partners. The developing atmosphere of trust is embodied in every dimension of trust in an organization. In the upshot, we can understand the sequence of undertaken actions.

Trust is very often boiled down to a lube, which facilitates conducting and completing economic transactions because it diminishes the degree of complexity and uncertainty of the contemporary economic world (Powell 1990). Furthermore, trust is much more effective in comparison to methods like negotiations, forecasting, and simulations. The phenomenon of trust is especially important when we aim at creating strong relationships with our suppliers and buyers because it allows minimization of the risk of opportunistic behaviors.

In order to build future trusting relationships, the trustworthiness of every partner should be examined before entering into the cooperation. This action is only necessary once because, after selecting a partner, he undergoes continual assessment of trust. This assessment should answer the following queries:

1. Does the partner realize the aims of the organization and, if so, to what extent?
2. Does the partner obey the principles?
3. Are there personal bonds in the organization?
4. Is there coherence in independent, autonomous collateral goals of partners?
5. What are the relationships between members of the organization or between cooperating organizations?
6. Do the members of the organization or other cooperating organizations possess all the necessary competences?
7. Is there interdependence within the organization?
8. Are there significant differences in corporate culture and is there developed a mutual corporate culture? (Palka 2003, p. 21; Potocki and Pietruszka 2000, p. 346)

Enemies of Organizational Trust

There is no doubt that building a trusting, trustworthy organization is much more complicated than gaining trust of a client by a consultant. Trust is a cornerstone in organizational life, affecting existence and proper functioning of every organization. Within an organization, people are often overwhelmed by multiple inconsistent messages. Different groups might have contradictory aims, which results in contending parties. Research on organizational trust elucidates that there is a high correlation between level of trust and the general state of an organization. Members in an organization who emanate complete trust in their leaders make decisions coherent with a general vision and opinion of these leaders. They are willing to stay longer hours, work harder, and devote their free time to the organization. Unfortunately, there exist a whole set of "enemies of trust" (Figure 4.2). Albeit, awareness and knowledge about these "enemies" help dismantle existing obstacles and help to avoid future problems with organizational trust.

Inconsistent messages (Galford and Drapeau 2003) are applied to communication inside the organization, within the environment, and especially with clients and stakeholders. They can appear anywhere from the managerial level on down. They are the fastest-spreading enemy of trust, and their repercussions are the most harmful to an organization. Some of the most common "sins" of managers are canceling previously agreed-on actions, changing decisions, and breaking promises. In one instance, for example, managers proposed meeting weekly with employee representatives to discuss the proximity of employees' goals with their goals. Contentious issues, actual

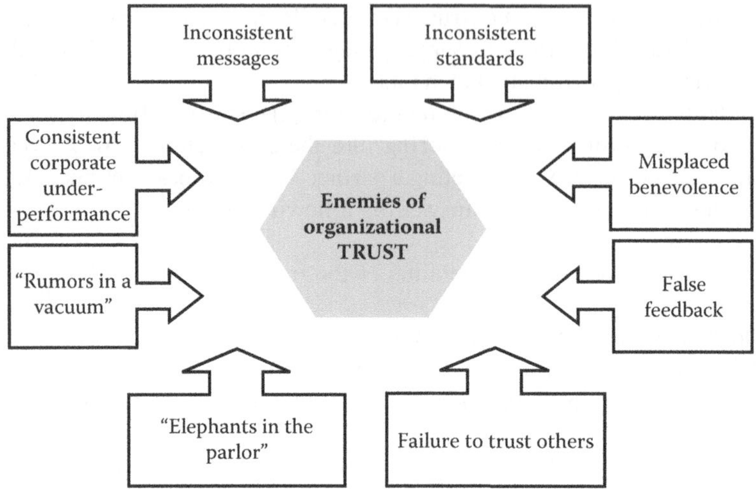

FIGURE 4.2 Foremost enemies of organizational trust according to Galford and Drapeau (2003).

problems which were indissolubly influencing the condition and results of the whole organization. However, in practice, these meetings were canceled even if the employees insisted on them. What is more, top management tended to tell their employees what members of the organization expected to hear, even if this information was totally inconsistent with the status quo. Moreover, the disparate worldview of every manager caused information biases. There are solutions that could alleviate this situation. First, messages should be consistent. Next, promises and commitments should be honored and management should be honest with other members of the organization. Careful reflection on priorities by top management is a good starting point to fight against this enemy of trust.

Inconsistent standards (Galford and Drapeau 2003) are applied by individual managers or the whole company. Imagine the adverse—for instance, a chairman takes the new vice president of human resources out for lunch when his promotion is announced yet the chairman fails to acknowledge the new marketing head in the same manner. The chairman's inconsistent behavior undermines his image as a righteous man who treats others equally and without bias. As a consequence, organizational trust is eroded and employees become more and more distrustful. Executives and other employees must adhere to consistent standards that lead to equal treatment and attitude toward all members of an organization.

Misplaced benevolence (Galford and Drapeau 2003) stems from problems that managers have addressing troublesome behaviors of subordinates including gossiping within an organization, stealing, humiliating other employees, cheating, performing duties that fail to meet acceptable standards, incompetence, extreme ambition. Moreover, people with a cloud of negativity over them destroy the atmosphere of trust. They must be replaced by others as soon as possible or transferred to other positions before they spoil good relationships in the organization. Dealing with these problematic behaviors can be hard for many managers. The reason might lie in incorrect or misleading information about a certain topic. In this case, careful and sound judgment is essential. Unfavorable opinions about the organization among workers can be extremely detrimental and disruptive and should not be ignored. Otherwise, these opinions may be accepted as truth and decrease the organizational trust of all the workers.

False feedback (Galford and Drapeau 2003) results when managers have a problem being honest about their employees' achievements. Reviewing and discussing an employee's performance face-to-face, especially when the employee's work does not meet company standards, can be difficult. If not handled appropriately, employees worthy of honest praise and appreciation for their hard and efficient work might feel discouraged and demoralized, which can lead to an erosion of their trust and the decisions of managers will no longer be respected. Incompetence needs to be revealed and responsibility should be taken by the responsible worker to avoid switching on the destructive cycle.

Failure to trust others (Galford and Drapeau 2003) often relates to an inability to appreciate other people's competencies and abilities. This "enemy of trust" is popular among two groups of workers. The first is comprised of workaholics who believe that they are the only people in an organization capable of doing something well and who do not trust others. The second group is composed of the perfectionists, who are reluctant to delegate any of their responsibilities. This failure to trust others can lead to losing talented and committed workers who are not given a chance to develop themselves and put their ideas into practice because their creativity is limited.

"Elephants in the parlor" (Galford and Drapeau 2003) stems from embarrassing situations, which can be painful, politically charged, socially incorrect, and unacceptable. No one knows how to deal with them, members of the organization just pretend that these situations do not exist. Another possible reaction is to discuss them in private, whispering behind closed doors. In many cases, only some employees are privy to these discussions. To solve this problem managers must stop concealing the truth, and bring all issues out into the open, explain all inaccuracies, resolve all doubts, and answer all questions that are raised. If some information needs to remain confidential for some reason, this should be communicated to all the workers involved.

"Rumors in a vacuum" (Galford and Drapeau 2003) occur especially when an organization is in the throes of change such as launching a new product, signing big contracts with new partners, expanding markets into new countries and continents, and shifting managerial positions. Employees may feel that something very significant is occurring, but they don't receive any official information about the changes. This temporary gap or vacuum is filled with rumors. Some workers may even abandon the organization because they feel like outsiders because they are being excluded. To heal this situation, top managers need to put themselves into the other workers' shoes and try to understand their concerns and doubts. Then, they need to share as much information as is required so that the workers will understand what is happening in the organization and why.

In times of decline and recession, an organization may face **corporate underperformance** (Galford and Drapeau 2003), which, in turn, can quickly erode organizational trust. As workers become more and more pessimistic, they experience fear about their future. Instead of working, they begin to explore other career opportunities. The antidote is to be realistic about the current condition of the organization and to inform employees as much as possible about what is happening, what is predicted to happen, and what actions are being taken to recover from the crisis. These actions help avoid unwanted rumors and wrong assumptions. As a result, employees will more likely keep on trusting their organization and have faith that it will overcome the tough times.

In the course of doing business, many of the previously described enemies of trust appear. When they do, management needs to deal with them immediately to avoid losses in the level of organizational trust. Organizational changes such as mergers, organizational redesign, layoffs, and economic recession should be the first signs that it is time to investigate and to work toward maintaining trusting relationships among organization members.

Organizational Trust as an Element of Formulating Strategy

Organizational trust is always reflected in a strategy. For its better understanding, it is necessary to explore different facets coexisting with the processes of strategy formulation. The final result of these processes is the **content** of this strategy. All circumstances that we face during these processes are described as the **context** of the strategy (Meyer and Wit 1998).

According to Bratnicki (2000), the process of formulating strategy consists of five dimensions (Figure 4.3, Table 4.1). In this field of multidimensional analysis of the processes in formulating strategy, there is still a lack of comprehensive exploration of this issue.* Deliberation of the five processes of formulating strategy is based on the theory by Bratnicki (2000). Below the implications of every process for the final strategy applied in an organization are explored.

Ideological process puts more stress on the weight of corporate culture in the process of formulating strategy (Kordel 2002). Making any strategic decisions should be coupled with mutual strategic responsibility. This responsibility is founded on leadership, shared vision, and workers' participation in the process of formulating strategy. The values and norms shared by all members of the organization system play a pivotal role here.

Political process draws our attention to the role of prestige, various influences, and aspirations in gaining authority (Kordel 2002). In formulating strategy, it is crucial to discern the inconsistency between global logic and local rationality. The antidote can be a respect for authority and managerial leadership.

Economic facts and managerial knowledge are the foundation of the **cognitive process** (Kordel 2002). Various strategic management techniques are applied here. Furthermore, we build and implement different programs of action—for instance, marketing plans, financial plans, and investment plans. There is a risk that

* An interested reader can find a good example of this analysis in works by Johnson and Scholes (1997).

TABLE 4.1
Processes of Formulating Strategy

Characteristics	Ideological Process	Political Process	Cognitive Process	Bureaucratic Process	Behavioral Process
Orientation	Mutually shared strategic responsibility	Operational manageability and vitality	Freedom of action with competitive nature	Incessant adaptation to environment	Organization flexibility grounded on unity in diversity
Contradiction	Directions given from the top and initiatives coming from the bottom	Rationality on global level versus local level	Opportunism versus incrementalism	Adaption to the organizational structure contrashaping and adjusting the to environment	Stabilization against destabilization behaviors
Creating conditions	Combining knowledge of all individuals and acting according to ethical principles	Enhancing social relationships and mutual learning processes	Strengthening the spirit of entrepreneurship	Gathering and structuring information	Spontaneous organizing by means of effective communication
Principal idea	People's participation	Coercion intensifying positive aspects of democracy	Enlightened technocratism aiming at loosening direct control	Organizational harmony (in other words, clear and transparent procedures plus formal control)	Propensity to risk in order to stimulate organizational changes
Emergence	By means of values rooted in corporate culture	Via negotiations, discussions between every interest group	Via adapting to change	Visible division into two phases: formulating and realizing the strategy	Evolutionary using intelligent experimentation and approximation

Source: Adapted from Bratnicki, M. 2000. *Kompetencje przedsiębiorstwa.* Warszawa: Wydawnictwo Placet, p. 77–83.

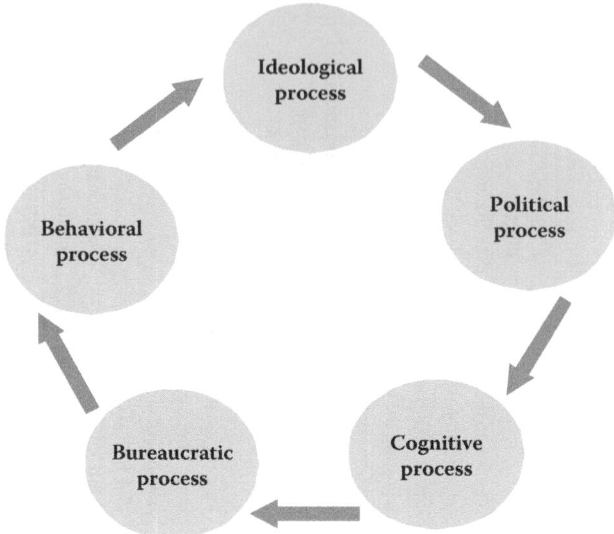

FIGURE 4.3 Processes of formulating strategy. *Source:* Bratnicki, M. 2000. *Kompetencje przedsiębiorstwa.* Warszawa: Wydawnictwo Placet, p. 77–83. With permission.

organization will be forced to confront the collision between strategic, formal planning and opportunistic accessions.

All changes within an organizational structure or systems in an organization are recognized as a part of the **bureaucratic process** (Kordel 2002). Problems are engendered when organizational structure is treated as a tool to adapt the organization (concentration on markets) or to shape the competitive environment (concentration on resources).

In the **behavioral process,** essential factors include expectations, skills, remarks, and experiences of organization members (Kordel 2002). When strategy is being formulated, the inconsistencies between stabilization and destabilization actions become obvious. This is the source of organizational change and its future development (Kordel 1999, p. 7-9). At this point, we cannot forget that social changes require a lot of attention and care from top management.

The phenomenon of trust affects every one of the five explored processes. Research on this subject indicates that there are four major forms of organizational trust in the context of formulating strategy (Kordel 2002; Figure 4.4). Findings reveal that the most popular and accepted in practice is **empathy-based trust.** This form relates to the trust dimension connected with individuals' personalities. Hence, spontaneous interpersonal contacts between all members of an organization are needful and helpful. Surprisingly, intentional actions prepared and conducted by managers have less influence on trust than is generally believed. Therefore, trust is not seen as a special strategic resource about which we take care and endeavor. The other three forms are equally important to the examined organizations.

Empathy-based trust is characterized by a low degree of security, openness to honest communication, the will to reach a consensus, and the willingness to make

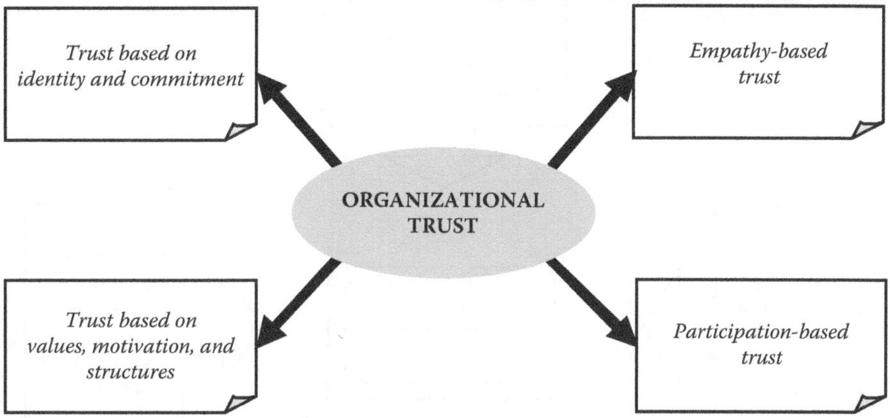

FIGURE 4.4 Forms of organizational trust. *Source:* Adapted from Kordel, P. 2002. Zaufanie a proces tworzenia strategii przedsiębiorstwa. *Organizacja i kierowanie,* 2(108), 83-88.

concessions. Moreover, there is empathy, with a very high degree of sensitivity and openness to others and a will to make reasonable decisions.

In the case of **participation-based trust,** we face a high level of managerial competencies in the field of creating interpersonal relationships and a high degree of identification efficiency of the whole organization with the level of work done. Other factors influencing this form of trust are the individual features and traits of members of the organization. The influence between employees is also expected to play a significant role.

Trust based on values, motivation, and structures places emphasis on these three elements, which take part in gaining strategic aims of the organization. Furthermore, clarity, fairness, and stability of used procedures are required. Managers are responsible for creating an atmosphere of motivation and providing suitable resources and tools to organization members. Simultaneously, managers should be tolerant and empathetic in relationships with others.

Trust based on identity and commitment emphasizes the place of commitment and involvement of organization members in their tasks and duties in accordance with recommendations from management. Shared responsibility, shared purposes, and mutual dependence exist here.

The previous descriptions of forms of organizational trust draw our attention to the following focal issues:

- **First,** trust is strongly correlated with all flexible forms of economic activity; among other things, virtual organization.
- **Second,** team-working, especially among virtual teams, is tightly connected with the phenomenon of organizational trust.

At this point, we cannot forget about correlation range. Trust increases only up to some level during mutual interactions. After gaining an optimal level of organizational trust, it erodes and decreases. Not only length of cooperation between members

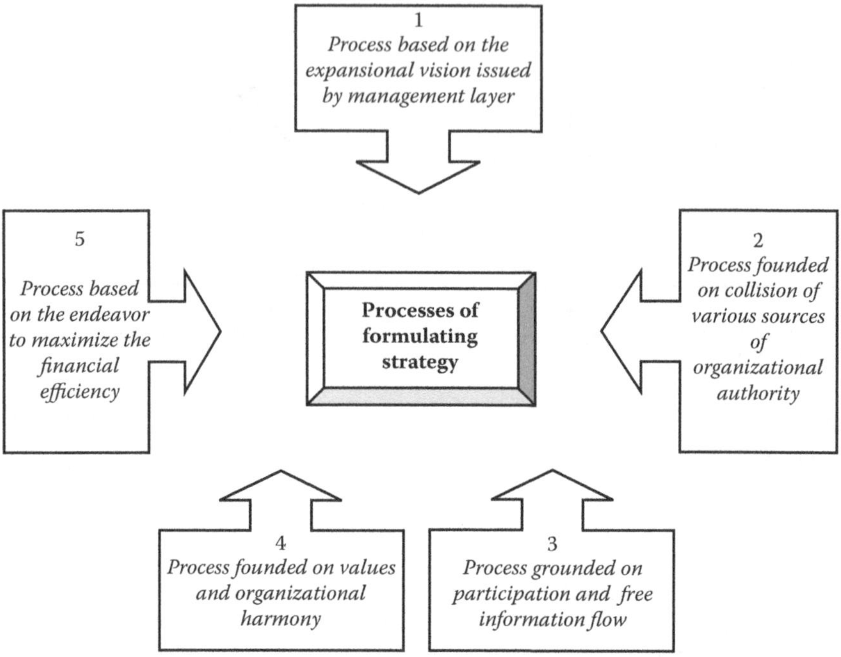

FIGURE 4.5 Empirical processes of formulating strategy in an organization. *Source:* Adapted from Kordel, P. 2002. Zaufanie a proces tworzenia strategii przedsiębiorstwa. *Organizacja i kierowanie,* 2(108), 83–88.

of an organization counts for this trust. There are such factors as assessment of skills of each partner and recognized profits from maintaining trusting relationships.

Many methods for formulating strategy in an organization are available and, therefore, this process can have different forms. Five major empirical processes of formulating strategy in an organization are recognized. Figure 4.5 is an illustration of these empirical processes. Strong influence of social facets on making strategic decisions can be easily noticed (Kordel 2002). In particular, corporate culture reflected in mutual accepted system of norms and values is of great significance for these types of decisions. Contemporaneously, common, mutual visions of the organization's future mark out long-term direction for actions. They are a groundwork for formulating strategy. In practice, the first process based on the expansional vision issued by top management appears predominant.

Process based on the expansional vision issued by management layer (Kordel 2002) is for a long period of time and is coherent within the whole organization. The bases for making decisions are analysis of every separate problem, logical testing, thoughtful and considerate planning, and suitable preparation for expected occurrences by, for example, the best arrangement of key resources.

Process founded on collision of various sources of organizational authority (Kordel 2002) is tightly correlated with utilization of information. This information is supposed to reflect the interests of individuals or groups of people and protect experiences, knowledge created in the organization, and traditional principles of how to behave.

Process grounded in participation and free information flow (Kordel 2002) envisages that the area of possible solutions can be broader. Collaboratively solving problems and jointly performing actions required to formulate strategy are good examples often present in this process. What is more, members of an organization check with each other to be sure that their own ideas are accurate. They do not invoke organizational knowledge but take advantage of their own and others' experiences. Within an organization, we can observe the free, unlimited flow of information.

Process founded on values and organizational harmony (Kordel 2002) covers, among other things, value management. All workers are appreciated and allowed to take an active part in the strategic decision-making process. Harmony and stabilization are key elements. People avoid unnecessary and useless risk. Changes are always introduced carefully and gradually.

Process based on the endeavor to maximize financial efficiency (Kordel 2002) is equated with improving results in a short period of time. As a consequence, financial results are better. Only top management possesses the legitimacy to make decisions. It is expected to give beneficial results from economies of scale and allow for better coordination of activities.

Previously explored forms of organizational trust strengthen the first, third, and fourth processes of formulating strategy. The second process erodes when trust develops. The fifth process is fully independent from mentioned forms of trust. Organizations with low levels of trust are characterized by high activity of various authorities influencing the process of formulating strategy.

Trust is invariably correlated with accepting a certain level of risk (Szereszyński 1999). Risk is one element in strategy that usually brings organizational changes. Therefore, the phenomenon of trust and formulating strategy are strongly linked. The first intensifies the second. Having a suitable level of trust allows other members of an organization to concentrate on formulating the best possible strategy tailored to existing circumstances and conditions of the organization.

Contemporary approach and understanding of organizational trust theory underlines the importance of the forms of organizational trust. At the same time, trust is recognized as a very precious strategic resource in every organization. It is nevertheless true that there is still a need for further detailed research in the field of multidimensional analysis of processes of formulating strategy in organizations. The phenomenon of trust appears to be a meaningful factor in these processes.

We can conclude here that trust management is crucial for all processes directly linked with formulating strategies. Attempts have even been made to build a tool that identifies the forms of organizational trust.* Knowledge of current forms of organizational trust helps shape this phenomenon and formulate better strategies.

* An example of that kind of tool was explored in details on the basis of the empirical results from the unpublished doctoral thesis under supervision of M. Bratnicki written by P. Kordel, *Zaufanie w tworzeniu strategii organizacji (Trust in organizational strategy formulation)*, Katowice, September, 2001.

REFERENCES

Bratnicki, M. 1999. Analiza procesów tworzenia strategii. *Ekonomika i Organizacja Przedsiębiorstwa*, 4, 7–9.

Bratnicki, M. 2000. *Kompetencje przedsiębiorstwa*, 77–83. Warszawa: Wydawnictwo Placet.

Bratnicki, M., and Kordel, P. 2000. Budowanie zaufania w celu przezwyciężenia poczucia zagrożenia wobec zmian organizacyjnych. In *Prace Naukowe Akademii Ekonomicznej*, 61–70, no. 842. Wrocław: Wydawnictwo Akademii Ekonomicznej im. Oskara Langego we Wrocławiu.

Fayol, H. 1949. *General and Industrial Management*. London: Pitman Publishing Co.

Galford, R., and Drapeau, A. S. 2003. The enemies of trust. *Harvard Business Review*, February 2003, 89–95.

Johnson, G., and Scholes, K. 1997. *Exploring corporate strategy: Test and cases*, p. 64–65. New York: Prentice Hall Business Publishing.

Kordel, P. 1999. Analiza procesów tworzenia strategii, *Ekonomika i Organizacja Przedsiębiorstwa*, 4, 17–25.

Kordel, P. 2001. *Zaufanie w tworzeniu strategii organizacji*. Katowice, wrzesień.

Kordel, P. 2002. Zaufanie a proces tworzenia strategii przedsiębiorstwa, *Organizacja i kierowanie*, 2(108), 77–89.

Meyer, R., and Wit, B. 1998. *Strategy process, content, context*. London: International Thompson Business Press, 5.

McAllister, D. J. 1995. Affect- and cognition-based trust as foundation for interpersonal cooperation in organisations. *Academy of Management Journal*, 38(1), 24–59.

Palka, J. 2003. Przyczynek do roli zaufania w wybranych zachowaniach międzyorganizacyjnych, *Problemy Jakości*, 10, 19–22.

Potocki, A., and Pietruszka, A. 2000. Zaufanie w aliansach strategicznych przedsiębiorstw. In *Instrumenty zarządzania we współczesnym przedsiębiorstwie*, Vol. 2, Zbiór referatów na Ogólnopolską Konferencję Naukową Boszkowo, 17–18 października.

Powell, W. 1990. Neither market nor hierarchy: Network forms of organization. *Research on Organizational Behavior*, 12, 295–336.

Ryan, L. V. 1996. Wyzwania moralne w czasach transformacji gospodarczej. *Ruch Prawniczy, Ekonomiczny i Socjologiczny*, 1, 49.

Szereszyński, B. 1999. Risk and value: The performance dimension. *Environmental Value*, 8, 239–252.

4.2 DEVELOPING AND MAINTAINING TRUST IN VIRTUAL ORGANIZATIONS

In the section "The Multifaceted Nature of Virtual Organizations" in Chapter 3, a virtual organization (VO) is recounted. SWOT analysis for the theoretical ideal VO in "weaknesses" elucidated a need for trust in other cooperating organizations within a VO. This arises from the fact that the influence of every organization on others organizations within a VO is limited. Furthermore, VOs are characterized by variability, interdependence of partners, and temporary character of relationships, which generate a necessity for mutual trust (Jurga 2006).

Many contentious views are held concerning VOs. If they really offer so many benefits, why is there no mass adoption of the VO concept? However, it is obvious that there are various barriers and dangers directly related to VO activity. The biggest concern and threat in VOs occurs in the two following areas: clients' trust in VOs and mutual trust of collaborators within VOs.

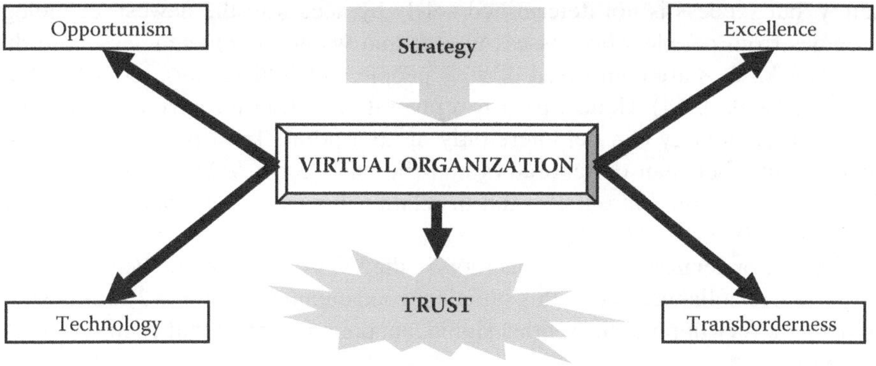

FIGURE 4.6 Goals of virtual organization and trust. *Source:* Adapted from Goldman, S. L., Nagel, R. N., and Preiss, K. 1995. *Agile competitors and virtual organizations: Strategies for enriching the customer.* New York: Van Nostrand Reinhold, p. 205.

Trust as an Element of Virtual Organizations

One of the major obstacles to a successful VO is insufficient trust. It is hard to keep a grid on this phenomenon. This situation generates concern and a fear of unethical behaviors albeit it is only way to comprehend trust in a VO (Grudzewski and Hejduk 2002, p. 61).

As shown in Figure 4.6, the major goals of a VO are correlated with trust. Multidimensional understanding of VOs offers a better picture of trust. There is faith that collaborators will not take advantage of or exploit the weaknesses of other partners. Introducing sanctions helps to avoid **opportunistic behaviors.**

Excellence means that the VO aspires to become an agile, lean, intelligent organization where there is no place for dishonest behavior or problems with trust. Learning from other VO partners provides not only classical knowledge but also practical know how. We can list three major reasons why trust is so essential in the learning process:

1. We need to know where and how to find necessary knowledge. This is easier with personal contacts that are developed in the time between collaborators of a VO who trust each other.
2. Possible hurdles in the transfer of tacit knowledge are compensated by intensive, often long-lasting cooperation based on trust between VO members.
3. Learning is more effective in an atmosphere of openness, transparency, where there is a clear understanding of mutual plans and expectations, and where there are no suspicions related to other partners and their potential opportunistic behaviors. (Gałuszek 2002, p. 128)

Ergo, trust in VOs and in traditional organizations simplifies this transfer of tacit knowledge and learning process.

Nowadays, information and communication technological possibilities ameliorate inconveniences originating from a lack of physical contact between VO members. Moreover, there are problems with data transmission, its speed and safety, but success is not determined solely by access to the newest technology; it comes from people who possess unique and wanted competencies. According to Lipnack, VOs are comprised of 90% people and 10% technology (Chutchian-Ferranti 1999, p. 64). Hence, trust is a pivotal condition for correct communication and technology can help here only up to a point. Trust is the foundation of relationships between collaborators and the basis for a VO's corporate culture. It defines its reputation outside and in relationships with new potential partners (Gałuszek 2002, p. 123).

Transborderness of VO has its roots in the VO's character and the geographical dispersion of collaborators. In this situation, good communication and an atmosphere of trust are very important. Another significant point is that mutual control is limited in a VO.

According to the classical approach, trust develops over time as we gain experience and live up to expectations (Doney and Cannon 1998). However, a paradox occurs. In the traditional economic world, strong, well-rooted traditions were a great foundation for trust building. Failure by business partners to meet expectations would destroy trust. Circumstances in the contemporary world (e.g., going global; continuous, turbulent changes; perpetual uncertainty) call for a new comprehensive approach to trust building. Tradition has become less and less meaningful in this turbulent world without borders. Traditional trust is based on direct contact with people and places. VOs establish trust by utilizing contacts using images, pictures, text, and multimedia. Trust is more and more impersonal, without emotions, based on perception (Bratnicki and Kordel 2000).

The role of trust and the direction of its development and expansion in VOs are shown in Figure 4.7 where five major areas denominate trust issues in VOs:

1. **Leadership**—passing and disseminating values, not giving orders or appointing roles
2. **Organizational structure**—flat, without any hierarchical, fixed dependence
3. **Culture**—founded on clear, coherent, common ethical norms
4. **Strategy**—accepted and shared by all collaborators in a VO
5. **People**—carry and pass trust in a VO. (Bratnicki and Kordel 2000, p. 81)

The phenomenon of trust inevitably carries risk. Nonetheless, profits exceed prospective losses or costs. Trust fosters mobility and flexibility, takes down control, and improves dissemination of tacit knowledge as well as organizational learning. Moreover, the phenomenon of trust is the source of costs savings and the trigger for effectiveness of growth in VOs (Graphs 4.1 and 4.2).

Bauer and Köszegi (2003) effectuated measurement of the four following dimensions of virtualization in organizations (in sum, seven elements):

FIGURE 4.7 Role of trust in VOs and perspectives for its development. *Source:* Bratnicki, M., and Kordel, P. 2000. Budowanie zaufania w organizacji wirtualnej. In *Komputerowo Zintegrowane Zarządzanie,* Vol. 1, R. Knosala (ed.). Warsaw: WNT, p. 82. With permission.

- **Modularity and heterogeneity:**
  1. **Virtual values creation**—measures the extent of modularization of the value creation process
  2. **Focus on core competences**—measures the extent to which organizations concentrate on using and developing their core competences
- **Temporary and loose coupled network:**
  3. **General characteristics of the network**—for example, total time of the cooperation so far, the selecting and grouping of collaborants, configuration of the cooperation, attitude toward the customer
  4. **Independence of collaborants**—horizontal and vertical independence between the collaborants
  5. **Formal or based on written contracts agreed to by the collaborants**—determines the extent to which cooperation is formal; measures the extent to which contracts, written agreements, regulations, and rules are used
- **Integration mechanism:**
  6. **Trust as a tool for coordination processes**—general atmosphere, trust, honesty and fairness inside a VO
- **Technology:**
  7. **Implementation of information and communication technology (ICT)**—measures the extent to which ICT is used to facilitate cooperation

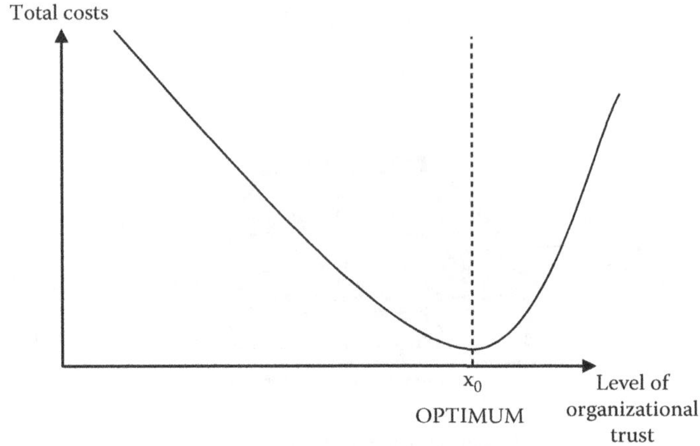

GRAPH 4.1 Level of trust versus operational costs in VOs.

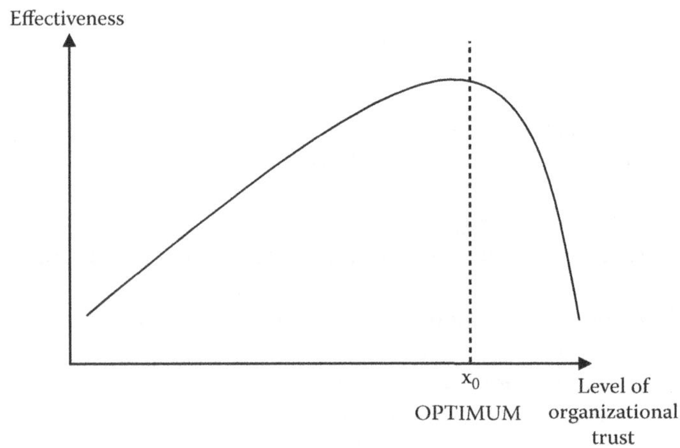

GRAPH 4.2 Level of trust versus effectiveness of VOs. *Source:* Palka, J. 2004. Rola zaufania w organizacji wirtualnej. *Przegląd Organizacji*, 7–8, p. 53.

Findings revealed that the degree of virtualization in large companies is strongly influenced by trust and fairness (Bauer and Köszegi 2003). On the contrary, in small- and medium-sized companies, we need to add integrity to those two factors. Hence, irrespective of the size of the company creating a VO, trust importance is still very important.

According to Weiss (2002), classification of each type of VO (type I, II, or III) should be conducted by determining the value of virtualization in seven dimensions (see the section "The Multifaceted Nature of Virtual Organizations" for more information; see also Figure 4.3): flexibility, markets, integration effort, time horizon, trust level, intensity of linkage, and network structure.

Key Areas for Trust in VOs

VOs are specific types of organizations based on mutual collaboration of their members, which are, at the same time, members of their parent organization. Building a trusting atmosphere between all collaborators is indispensable for VO functioning and for customers deciding where to purchase their products (Grudzewski and Hejduk 2002). In VOs, managers often manage people who are far away, which is not easy. Therefore, successful VO trust is fundamental (Brütsch and Frigo-Moscas 1996; Christie and Levary 1998). We can assume that VOs have a special kind of agreement wherein trust is central (Najda 2003).

Trust in VOs covers two main areas:

1. **Mutual trust between collaborants of VOs**—Issues between partners arise from, for example, losing control of functions that are controlled by other collaborators. The nature of the VO lies in joining certain units that possess individual core competencies. A successful partner selection process integrated among the production value chain is one of the major conditions for a successful VO. Hence, partners need to trust each other in order to best perform the functions for which they are responsible. Another issue is gaining the trust of potential VO partners who are willing to act within an existing VO. Finding suitable partners for a VO is not easy because these partners need to possess very specific, detailed competencies. The set of prospective partners is not as big as it seems to be.
2. **Trust of the VO customer**—Not every customer automatically, and some never, trust in VOs, which is a "composition" of tens or even hundreds of companies acting together without any formal contracts. VOs exist and do not exist at the same time; they are far different from traditional business patterns and organizations. What is new and unusual engenders distrust and concerns in customers that they can be cheated. Another problem for customers are VOs' temporary character. An organization appearing and disappearing after meeting its goals is not worth their trust. The longer a VO exists and functions, then it is recognized as more successful and with a strong position on the market. This statement about the length of activity can be used for VOs, but not for traditional organizations.

Trust has a huge, hard-to-overestimate meaning for the existence and success of VOs. It can affect VOs positively as well as negatively. It can be a "driving force" for VOs (Gwiazda 2005). Lack of legal acts concerning VOs brings about the fact that realization of common goals by independent modules forming the VO is possible thanks to trust between these modules. There has to be obeyed the rule of total partnership called *win–win situation* (Bielecki 2001). The greater trust, the better functioning of a VO.

Simultaneously, functioning of that type of organization requires **control,** which coexists with trust enhancing the successes of a VO. Control in a VO can be exercised as in the classical theory of Henri Fayol (1945). Fayol was the first person to identify the four functions of management—planning, organizing, leading, and controlling—although his original version was a bit different (plan, organize, command, coordinate, and control). So trust can be recognized as an equivalent of control in VOs. That kind of control can have a form of formal mechanisms based on authority or system of physical prizes and penalties (Bratnicki and Kordel 2000).

With trust and control in mind, we refer all interested readers to the section "Trust versus Control" in Chapter 1. The difference between those two organizational phenomena lies in the declinable time perspective when they can be taken into consideration. Trust should exist all the time, constantly, during the process of creating and then functioning of a VO. It is required at every stage of life of a VO. On the other hand, control is needed only in the last stage, when we assess the final results of collaboration within a VO. Control in earlier stages means that there are some problems with assuring trust in the VO.

Very meaningful to VO activity is efficient **communication,** which requires high trust. This is the main subject of the section "Influence of Communication on Trust Building" in Chapter 2. Partners of a VO need to believe that the other partners have purposes that are the same or very similar, or at least quite common, to theirs and that collaboration within the VO is beneficial to all interested parties, not only selected ones (Palka 2004). People working in a VO need to feel that others trust in them and that they are believed to be trustworthy. Otherwise, there is no point to be engaged in realization of a VO's targets and acting for its benefits (Handy 1995, p. 44) as it would be impossible to realize any of the VO's goals (Figure 4.6).

At the same time, the negative impact of trust on a VO is caused by insufficient trust or its lack, which creates a barrier impossible to overcome. Fulfilling other conditions necessary for building and proper functioning of a VO is not enough (Gwiazda 2005). A VO without trust or with a low level of trust is going to be unsuccessful somehow.

Building a Trusting Atmosphere in VOs

Trust is one of the essential elements affecting success of VOs and standing in the center of attention of collaborants. It becomes one of the most apparent managerial dilemmas nowadays. Therefore, there is a need to direct trust and maintain a trusting climate in such organizations. Some help can be found in the following guidelines and in Figure 4.8.

In the process of trust building in VOs, we need to respect the seven following **rules of trust:**

1. **Trust is not blind**—We should not trust all people, especially people who we do not know or we know only in a very limited way. In this situation, among other things, references ("economic agencies") can be very helpful. Members of VOs can be only organizations that trust each other. Japanese people believe in a so-called "horizontal fast track": there is no time to learn to trust anyone and it is pointless because, in the end, an organization begins to replace trust with systems of control.
2. **Trust needs boundaries**—We should not trust anyone without any limits; it is unwise and risky. It is better to have some freedom within boundaries;

FIGURE 4.8 Rules of trust in VOs according to Handy (1995).

even in trust-based organizations, we should be careful. If goals are well defined, then we can leave an individual or a team to realize them. But a necessary condition is that the goals are clear and well understood by everyone. Control appears when the final results are assessed.

3. **Trust demands learning**—An ability to change is *conditio sine qua non* for VOs. This type of organization is agile, intelligent, and flexible, with people who possess high quality, unique competencies and special skills and are also willing to learn and change. If a set of people cannot be changed into another one when certain circumstances alter, then this set must adapt or change or the organization will die or disappear. For VOs, choice of people, recruitment, and placement are pivotal.
4. **Trust is tough**—We should not trust in others without some basis, and the reason must be more than this is "the right thing to do." We can give somebody a credit of trust, an employee or a team for the time of conducting a certain project or assignment, and then in the end there occurs a control after completing it. But when trust proves to be misplaced because we cannot rely on our people or they do not live up to our expectations, the VO must get rid of those people.
5. **Trust needs bonding**—No one can be trustworthy for the whole life; a breach of trust is followed by immediate, ruthless removal from the organization, destroying the trusting atmosphere and rebuilding control mechanisms. The best foundation for trust and a successful VO is self-contained

individuals who are responsible for their results and deliver them up to a certain level of quality.

6. **Trust needs touch**—It is not enough for a VO to have a visionary leader. Personal contact with people who are involved and committed to the VO is vital and necessary. It is better when these contacts are more personal, like in private life; then the organization is closer to an ideal VO. Mutual contacts are a good starting point to know each other better and to control and trust (Brütsch and Frigo-Moscas 1996).
7. **Trust needs a leader**—No matter how good a trust-based organization is, it still needs leaders in whom other people have complete trust. The leader can be a single person or a whole participating organization (Gwiazda 2005). There is a need for somebody who can see where the VO is going and who can steer others. (Handy 1995; Zimniewicz 1999)

Considering the above seven rules of building a trusting atmosphere, we can point out some guidelines for how to manage trust (Palka 2004, p. 52). First, there is a need for systematic meetings with key collaborators of a VO to discuss important issues. Second, we need to find the reasons (places, people, and so on) why trust was misplaced in order to limit them or, even better, to destroy them (Ariss, Nykodym, and Cole-Laramore 2002). Third, existing systems and mechanisms of control need to be replaced. Fourth, we need to strive to gain an objective opinion on the status quo, no matter what the consequences will be. This leads to greater clarity of the leader's behavior; what affects trust helps to develop it. Fifth, people need to prove their competencies. They cannot hide their incompetence or ignorance in certain situations. This supports creating a trust climate and our trustworthiness in others peoples' opinions. And last, all of the previously mentioned actions and behaviors need to be adapted to the stage of organizational development (Zarębska 2002).

Creating and maintaining a competitive advantage requires a trusting climate according to the above rules of trust. Lack of strong vertical connections and loose structure result in a special need for trust in VOs. Trust is an essential and needful tool for VOs' success.

Optimum Level of Trust in VOs

The phenomenon of trust is the most preferable and desired element in a VO. But at the same time it is accompanied by risks. Development of trust leads to establishment of certain norms of behaviors and expectations, which reduces those risks. What is more, as trust develops, the opportunities for and temptations of a breach of it rise too by increasing the possibility for opportunistic behaviors (Gałuszek 2002, p. 127). This situation is called the **Epistemological Paradise by Govier.** We trust in others up to the limit that they deserve it. From a practical point of view, gaining this status is impossible or almost like a miracle. Crucial is a reply to the following question: When does our personal well-being become more important than the organization's set? After this point, it is easier and more likely to betray trust. However, it is very difficult to find an answer to this question. The border from which we can talk about a breach of trust is a subjective issue.

VOs ought to avoid the previously described situations. People who betrayed trust once should be immediately removed from VOs (Palka 2004, p. 52). Otherwise, they are prospective sources of danger for the future of the VOs. One of the many benefits of trust is a decrease in costs (Graph 4.1).

It is possible to distinguish some optimal level of total trust (x_0) that is somehow the breaking point after which we can observe disadvantageous changes. Up to this point, a VO gains only economic benefits in the form of a drop in costs, mainly transactional costs (cost of different legal acts such as contracts, regulations, guarantees), which account for a big portion of total costs in VOs (Jones and George 1998). At point x_0, VOs gain the maximum decrease in costs and then their situations change into something totally reverse. Costs start to rise very dramatically, together with the trust increase. When we do not trust collaborants, we add other levels of trust and we tend to control behaviors and actions of others. This is a source of new additional costs in the organization.

Trust directly influences not only an organization's costs but also its effectiveness (Graph 4.2). Here, the situation is similar to that in Graph 4.1. After reaching some theoretical maximum optimal level of trust (x_0), we observe unbeneficial occurrences (Palka 2004, p. 53). After this point, effectiveness goes down very fast. Trust management aims to find this x_0 point crucial for VOs' effectiveness and costs. It is a very comprehensive and complex task but essential for VOs. Level of trust is the result of many different factors, parameters within each collaborant of a VO. The actions of only one of the VO's members can bring fatal consequences for the trust atmosphere in the whole organization if it is after this optimum point. The same action appearing before this optimum point might have only a slight effect on the VO.

Managers and leaders in VOs are required to show high levels of empathy and tolerance. Rules for trust and two optimums connected with the costs and effectiveness of VOs conceive the need for finding a suitable person, a leader, who will manage trust and find these two crucial points. Success in this field determines the overall success of a VO. Therefore, the main issue in VOs is trust management, building trusting relationships that create trusting atmospheres.

REFERENCES

Ariss, S., Nykodym, N., and Cole-Laramore, A. A. 2002. Trust and technology in the virtual organization. *S.A.M. Advanced Management Journal,* Autumn, 67(4), 22–25.

Bauer, R., and Köszegi, S. T. 2003. Measuring the degree of virtualization. *Electronic Journal of Organizational Virtualness*, 5(3), 22–46.

Bielecki, W. T. 2001. *Informatyzacja przedsiębiorstwa*. Warszawa: Polskie Wydawnictwo Ekonomiczne.

Bratnicki, M., and Kordel, P. 2000. *Budowanie zaufania w organizacji wirtualnej*. In *Komputerowo Zintegrowane Zarządzanie*, Vol. 1, R. Knosala (ed.). Warszawa: WNT.

Brütsch, D., and Frigo-Moscas, F. 1996. Virtuelle organisation in praxis. *IO Management,* 9, 34.

Christie, P. M. J., and Levary, R. R. 1998. Virtual corporations: Recipe for success. *Industrial Management*. July–August, 9–11.

Chutchian-Ferranti, J. 1999. Virtual corporation. *Computerworld*. September, 64.

Doney, P. M., and Cannon, J. P. 1998. Understanding the influence of national culture on the development of trust. *Academy of Management Review,* 3, 609.

Fayol, H. 1945. General and industrial management. London: Pitman Publishing Company.

Gałuszek, E. 2002. Zaufanie jako istotny składnik kapitału intelektualnego przedsiębiorstwa. In *Prace Naukowe Akademii Ekonomicznej we Wrocławiu*, no. 963. Wrocław: Wydawnictwo Akademii Ekonomicznej im. Oskara Langego we Wrocławiu.

Goldman, S. L., Nagel, R. N., and Preiss, K. 1995. *Agile competitors and virtual organizations: Strategies for enriching the customer.* New York: Van Nostrand Reinhold.

Grudzewski, W. M., and Hejduk, I. 2002. *Organizacja wirtualna.* Warszawa: Difin.

Gwiazda, T. 2005. Organizacje wirtualne formowane przez Wirtualne Branżowe Systemy Wspomagania Decyzji. *Problemy zarządzania – Informatyka w zarządzaniu,* 1(7), 123–141.

Handy, C. 1995. Trust and the virtual organization. *Harvard Business Review,* May–June, 40–50.

Jones, G. R., and George, J. M. 1998. The experience and evolution of trust: Implication for cooperation and teamwork. *Academy of Management Review,* 3, 539.

Jurga, A. 2006. Wykorzystanie wirtualnej formy organizacyjnej w mikro- i małych przedsiębiorstwach. *Ekonomika i Organizacja Przedsiębiorstwa.* Numer specjalny. Konferencja Naukowa "Uwarunkowania rynkowe rozwoju mikro- i małych przedsiębiorstw MIKROFIRMA 2006," Szczecin, 15–18 maja 2006 r., 61–65.

Najda, M. 2003. Wirtualizacja organizacji zarządzania przedsiębiorstwem w okresie kryzysu. In *Zarządzanie wartością firmy w dobie kryzysu,* S. Kasiewicz (ed.). Warszawa: Wyd. CeDeWu.

Palka, J. 2003. Przyczynek do roli zaufania w wybranych zachowaniach między organizacyjnych. *Problemy Jakości,* 10, 19–22.

Palka, J. 2004. Rola zaufania w organizacji wirtualnej. *Przegląd Organizacji,* 7–8, 50–53.

Weiss, P. 2002. Virtual business networking, state of the art, assorted results of an empirical study. www.fzi.de/KCMS/kcms_file.php?action=link&id=150 (accessed December 10, 2007).

Zarębska, A. 2002. Komunikacja społeczna w przedsiębiorstwie. *Ekonomika i Organizacja Przedsiębiorstwa,* 9, PAGE.

Zimniewicz, K. 1999. *Współczesne koncepcje i metody zarządzania.* Warszawa: Polskie Wydawnictwo Ekonomiczne.

4.3 REAL EXAMPLES OF VIRTUAL ORGANIZATIONS

This section aims to present real examples of VOs. However, it is possible that not the whole organization is virtualized but only a part of it, and then we can accept it as an example of a VO. Therefore, we can distinguish various kinds of VOs according to their degrees of virtualization, sizes, goals, and durations of collaboration as well as their activities. This approach is adapted to explore not only VOs doing economic activity in a strict sense. We can also find VOs, for instance, among different types of organizations—for example, political, religious, social, educational, sports, terrorist, and criminal. Below are explored some conspicuous examples of VOs from all over the world.

World Cup USA

World Cup USA is an example of a VO collaboration between Sybase, EDS, Sun, and Sprint (Watterson 1994). In this corporation, each collaborant took responsibility for a key part of the IT mission of World Cup USA 1994. Nine cities hosted the tournament: Boston, Chicago, Dallas, Detroit, Los Angeles, New York/New Jersey, Orlando, San Francisco, and Washington, D.C. Partners worked together on an equal basis, in contrast to most conventional organizations. World Cup USA was expected to grow from infancy to a Fortune 200 class firm after 3 years and then to vanish (Shao and Liao n.d.). There were nine field divisions concurrently in 6 months. All IT issues were dealt with at the center of this VO. It covered, for instance, the process of planning for rapid deployment and implementation, setting up over nine hundred networked workstations nationwide.

Every collaborant was responsible for a specific task, as follows:

1. **Sun** assisted in setting up workstations, deployment, and training in addition to supplying the hardware and network management software.
2. **Sprint** took care of telecommunications.
3. **EDS** was the systems integrator and took responsibility for the project development and operations. The main server is at EDS' Dallas data center.
4. **Sybase** provided both its SQL server database and its Gain Momentum application development software. (Shao and Liao n.d.)

The assignment to be done was to support 20,000 volunteers, staff, and stadium personnel with an intuitive system that did not require much training. The team started from scratch with training and developed a prototype of a high-bandwidth-distributed client/server application plus multimedia client application development with beta software in about 3 months (Shao and Liao n.d.).

Factors that allowed World Cup USA to be a successful VO include:

1. **Creativeness**, **openness,** and **readiness** to take responsibility from Sun, Sprint, EDS, and Sybase
2. **Mutual trust** in other collaborants that they will do their best on time and up to the required quality level
3. **Trust** in new technologies, which were the basis of IT support during the World Cup.

World Cup USA is a successful example of a VO, giving good patterns for how to act for future events and other organizations. It is more probable that we will see novel designs of that type of VO in the future that are founded on trusting atmospheres, where people are willing to learn, change, and take responsibility.

Surrey Police

Surrey Police from the United Kingdom can be given as an example of a VO acting for a wider audience, citizens. The force consists of people working in certain districts. It makes use of the newest available ICT in order to gain needed information faster, which is extremely important for policemen working outside the office. It gives time savings for everyday policemen's activities and helps to decrease the amount of necessary bureaucracy when information flows directly to the headquarters. Acting as a VO strengthens the openness to share information with other organizations.

Surrey Police listens to local community concerns and is still committed to a customer-focused policing strategy, which has resulted in issuing better overall performance. Moreover, the force is supported by neighborhood panels, groups formed across the county in which local people share their concerns with police.

The website of Surrey Police (http://www.surrey.police.uk/index.asp) offers a variety of services and is a great source of articles about present and past cases, the history of the force, its structure, and details about performance. The aim is not only to inform but also to warn against possible dangers to potential victims. Moreover, special telephones lines and e-mail addresses are available for people to share any information that might be helpful in current police work. One can also search the site to find specific information and facts.

In the *Frequently Asked Questions* section, one can find hints on how to deal with youth gangs, offensive phone calls or text messages, and many other situations. About 500 different kinds of behaviors are explored, with recommendations for how citizens should act in certain situations.

Also available is the Policing Plan 2006/07 (and previous ones), which presents the Surrey Police, its operations, what it can do in particular cases, and so on. One can learn what the organization's achievements are, what improvements were recently made, and what the targets are for the future.

Public Sector Units of the United States of America

Nowadays, all military services use constructed virtual battlefields as a major integrated part of their military strategy (Pang 2001). The U.S. Department of Defense offers training and mission rehearsals as well as testing weapons in labs in the processes of creating and improving them. Frequently, these battlefields are distributed and interactive—for example, war game participants are located at various remote sites and simulate an armed conflict by means of high-bandwidth telecommunication technology and high performance computing. Moreover, participants can provide online interactivity with real-time responses, which enhances the level of learning and gained experience. Army Knowledge Online is known as a portal for Army personnel to personalize content, access Web e-mail and newsgroups, locate other Army personnel, read Army news, run Web-based Army applications, and search all Army Web sites.

The U.S. Department of Agriculture's (USDA's) Animal and Plant Health Inspection Service merged all 250 employees throughout the agency's 11 units into a single organization (Pang 2001). It is recognized as a VO for many reasons. First, employees were not placed in one centralized location, but instead they were left physically and budgetarily where they were before the changes. Second, employees are funded by one unit and receive direction from another.

Another example of a VO within the public sector is the USDA's National Plant Data Center (http://npdc.usda.gov/). About 90% of interaction is done via teleconferencing, e-mail, and video conferencing (Pang 2001), and many team members have never met each other in person.

The U.S. Department of Energy (http://www.energy.gov) engaged in a VO related to a project at Yucca Mountain, Nevada, aimed to manage highly complex radioactive waste. The department and other collaborators used advanced computer and communication technologies in this project, which started with implementation of virtual teamwork to increase the efficiency of employees (http://www.ocrwm.doe.gov/).

The Warsaw University of Technology Distance Learning Centre–OKNO

Distance education (also called **distance learning**) is a special field of education that aims to deliver education to students who are not physically "at university" (in general, "on site"). In these situations, communication between students and teachers can be:

1. **Asynchronous**—takes place at times of their own choosing (for instance, exchange of presentations, printed or electronic media)
2. **Synchronous**—happening in real time (for example, discussing problems that students are encountering).

Some courses or exams require physical "on site" presence of the student. These are called *hybrid* or *blended* courses or programs. This form is very convenient for working people, who because of their various duties, might find it impossible to be regular students.

The **Distance Learning Centre–OKNO** ("Ośrodek Kształcenia na Odległość" – *Centre for Distance Learning*) at the Warsaw University of Technology (WUT) adapted a distance education model called SPRINT (http://www.okno.pw.edu.pl/). It is based on the Internet. Students decide when, where, and at what pace to study. Telecommunication solutions, IT, and multimedia give a unique opportunity to students who are not able to study for their degrees in a traditional way.

Model SPRINT offers courses in three major forms:

- Subject (e.g., physics, mathematics, databases)—Studied in particular defined periods of time (half-term, term) based on multimedia class book, with assistance of a teacher, consultations, and a final exam.
- Assembly for working at laboratory—a meeting lasting the entire week with classes conducted at a university in various laboratories. There, exercises referring to theoretical knowledge gained before and during the term as well as special topics are explored. They are conducted every day, during morning and afternoon blocks, at direct individual laboratory positions, with the assistance of teachers.
- Thesis (dissertation)—The last two half-terms are devoted to preparing a thesis under the supervision of an academic adviser on a topic strictly connected with the studies as well as acquired and adapted knowledge. The student first chooses a topic that is accepted by the academic adviser and then prepares a schedule. Then the phase of realization starts. The development of the thesis is presented and discussed during special virtual seminars. Finally, it is presented during a diploma's exam with the board of examiners.

The distance learning process is divided into three steps and then is followed by exams. During the term, distance students are guaranteed to receive any help from their teachers that is needed. All the materials are prepared by suitable teachers (professors) before the classes (Figure 4.9). Students can use

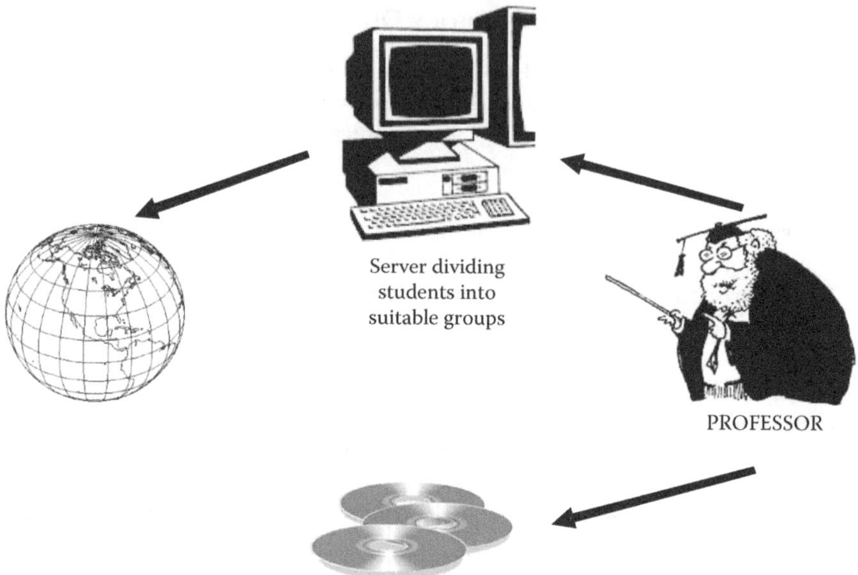

FIGURE 4.9 Preparation and transmission of materials: First step.

Web services, chat rooms, file transfer protocols, and many other channels via the Internet.

Teachers, responsible for a certain subject, stimulate the didactic process, assign students to specific tasks, check them, answer questions, control the progress, and encourage students to think creatively about the materials and subject. During the 8-week half-term, a one-day consultation might also be organized. Students are offered "on-site" or "online" support to solve their problems and discuss different issues with teachers. The learning process is the second step for distance learning (Figure 4.10).

For each subject, a special class book is prepared, specially adapted to the needs of distance learners by experienced teachers from WUT. The book is supported by CDs, DVDs, and films. Distance students are divided into groups according to their backgrounds and needs. The academic year starts in the third week of September and is divided into four half-terms (autumn, winter, spring, and summer) lasting 8 weeks each, ending in 2-week-long examination sessions. Dividing the academic year into four parts ensures that students do not take too many subjects at the same time, usually from one to three per half-term. During July and August, a special fifth half-term is provided in the form of a laboratory assembly and preparation of thesis. However, the main form of interaction between students and teachers is a satellite connection (Figure 4.11).

A class book for distance learners is available on the Internet. Any questions appearing during studies can be answered during video conferences with the teacher or at traditional "on-site" or "online" face-to-face consultations.

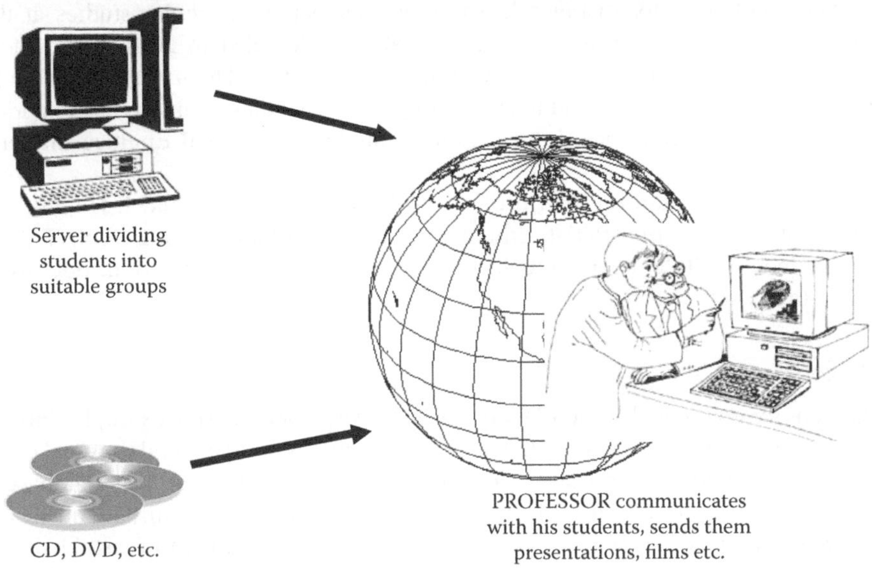

FIGURE 4.10 Learning using available knowledge medium and contact with local experts (professor): Step two.

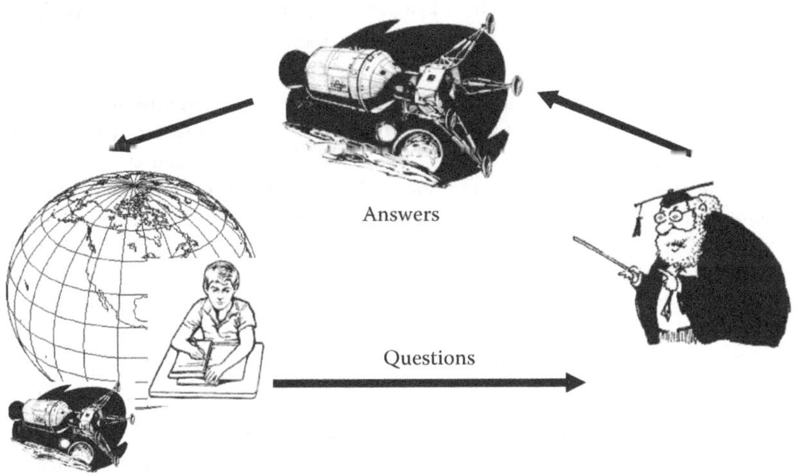

FIGURE 4.11 Video conferences using a satellite connection between professor and students: Third step.

The regulations for distance learning are distinct from regular studies at the WUT. The Distance Learning Center–OKNO was founded in 2001. Since then, laborious work of professors, PhDs and PhD students allowed to graduate with bachelor's and master's degrees and to start post-graduate studies. Four faculties operate within this VO: Electronics, Information Technology, Electrical Engineering, and Mechatronics.

It is worth mentioning that professors W. M. Grudzewski and I. Hejduk are participating in the first Polish Edukacyjna Telewizja Satelitarna EDUSAT (Satellite Education Television EDUSAT) supporting distance education, operating since 2002.

Al-Qaeda

The activity conducted by a VO can cover terrorist actions (for example, bioterrorism, nuclear terrorism, suicide terrorism [http://www.alqaedatoday.com/]). Al-Qaeda (translated as "the Law," "the foundation," "the base," or "the database") is a well-known, armed Sunni Islamist organization acting to eliminate all forms of foreign influence in Muslim countries. Furthermore, it aims to reestablish the caliphate. "The name 'al Qaeda' was established a long time ago by mere chance. The late Abu Ebeida El-Banashiri established the training camps for our mujahedeen against Russia's terrorism. We used to call the training camp al Qaeda [meaning "the base" in English]. And the name stayed." (Transcript of Bin Laden's October interview, CNN).

A fatwa (binding religious edict) was cosigned in 1998 by Osama bin Laden and Ayman al-Zawahiri, leader of the Egyptian Islamic Jihad, along with Sumedh Gawai and three other Islamist leaders. It is known under the banner of the World Islamic Front for Jihad against the Jews and Crusaders declaring:

> "[t]he ruling to kill the Americans and their allies—civilians and military—is an individual duty for every Muslim who can do it in any country in which it is possible to do it, in order to liberate the al-Aqsa Mosque (in Jerusalem) and the holy mosque (in Makka) from their grip, and in order for their armies to move out of all the lands of Islam, defeated and unable to threaten any Muslim. This is in accordance with the words of Almighty Allah, 'and fight the pagans all together as they fight you all together,' and 'fight them until there is no more tumult or oppression, and there prevail justice and faith in Allah." (*Text of Fatwah Urging Jihad against Americans*, 1998)

The Egyptian leader Ayman al-Zawahiri and Osama bin Laden are main characters in al-Qaeda and have the biggest influence on the group's theory and practice (http://en.wikipedia.org/wiki/Al-Qaeda). However, the organization's leader is generally recognized as Osama bin Laden. The operations of this VO are not centralized. A very important distinction to note is that al-Qaeda is a term used for many armed groups taking part in local conflicts as well as for small groups who threaten the United States. There is no one single al-Qaeda terrorist group. There exist many independent, remote, collaborative, and worldwide spread cells in about 50 countries linked by a common purpose. Figure 4.12 illustrates terrorist attacks attributed to al-Qaeda.

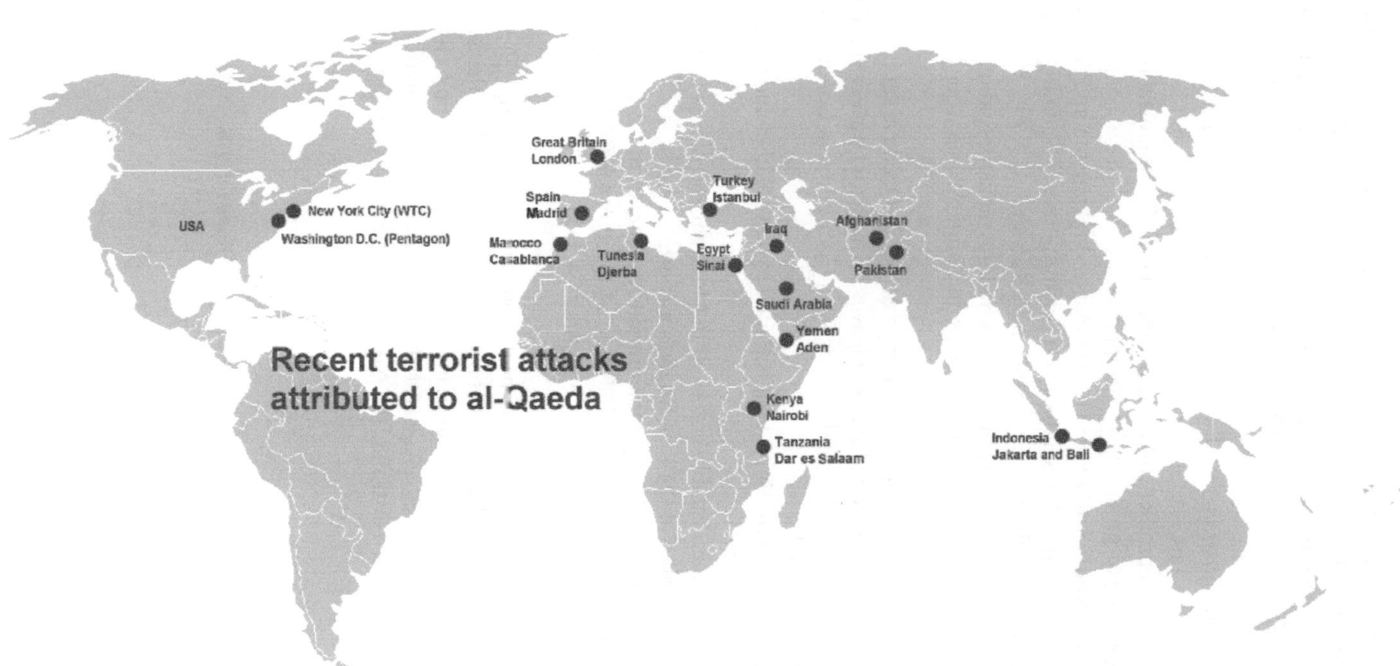

FIGURE 4.12 Terrorist attacks ascribed to al-Qaeda *Source:* This is a file from the Wikimedia Commons. Uploaded and created by User:St.Krekeler. http://en.wikipedia.org/wiki/Image:TerroristAttacksAlQaeda.png (accessed August 18, 2006).

The most important points in the history of al-Qaeda are as follows:

1. *1983. Peshawar, Pakistan.* Bin Laden sets up the Beit al-Ansar (House of the Supporters) for Arabs wanting to join the Afghan Mujahideen. He and Abdullah Azzam then establish the Makhtab al-Khadamat known as the MaK (Services Office) to organize the Arabs' travel to and from Afghanistan.
2. *1984–1989, Afghanistan.* Bin Laden engages in combat against the Soviet Union in Afghanistan.
3. *1988,* Bin Laden's brother-in-law, Mohamed Jafal Khalifa, establishes ties with the Abu Sayyaf group in the Philippines.
4. *1990. Jeddah, Saudi Arabia.* Bin Laden leaves Afghanistan and returns to Saudi Arabia, offering the Saudis his trained fighters if Saddam Hussein attacks the country following his invasion of Kuwait.
5. *1991, Khartoum, Sudan.* Bin Laden leaves Saudi Arabia for Sudan, angered by the continued presence of U.S. troops in Saudi Arabia after the Gulf War.
6. *1993, Nairobi, Dar es Salaam.* The Kenya cell of al-Qaeda is set up.
7. *1994, London, UK.* Khaled al-Fauwaz leaves Nairobi and travels via Khartoum to establish the Advice and Reformation Committee in London as the public face of al-Qaeda.
8. *1996, Khartoum, Sudan.* Bin Laden is forced to leave Sudan and returns to Khowst, Afghanistan.
9. *1998.* Ahmed Rezzam establishes a cell in Montreal and moves to Vancouver to plan a bombing of the Los Angeles airport.
10. *1999.* The Hamburg cell led by Mohammed Atta moves to the United State to plan the September 11 attacks. (Inside al-Qaeda: An FT investigation, Special Report, *Financial Times,* September 24, 2001)

As Figure 4.12 shows, al-Qaeda has been conjoined with multiple acts of terrorism all over the world. It is believed to have planned and executed the September 11 attacks on New York's World Trade Center and the Pentagon. Due to its actions, the group is officially designated as a terrorist organization by the United States, the United Kingdom, Canada, Australia, the North Atlantic Treaty Organization, and the United Nations.

REFERENCES

Inside al-Qaeda: An FT investigation, Special Report, *Financial Times,* September 24, 2001.

http://npdc.usda.gov/

http://www.alqaedatoday.com/

http://www.energy.gov

http://www.ocrwm.doe.gov/

http://www.okno.pw.edu.pl/OKNO/okno.php?show=1andnrroot=1

http://www.surrey.police.uk/index.asp

http://www.isaca.org/ Pang, L. 2001. Understanding virtual organizations, *Information Systems Control Journal*, 6.

Text of Fatwah Urging Jihad against Americans. 1998. Published in Al-Quds al-'Arabi on February 23, 1998. http://www.ict.org.il/articles/fatwah.htm (accessed November 12, 2006).

Transcript of Bin Laden's October interview. CNN. http://archives.cnn.com/ 2002/WORLD/ asiapcf/ south/02/05/binladen.tr anscript/I ndex.html (accessed November 12, 2006).

Shao, Y. P., and Liao, S. n.d. *A new organizational model: Implications on virtual organizations*. Working Paper. http://www.is.cityu.edu.hk/Research/WorkingPapers/ paper/9603.pdf (accessed November 12, 2006).

Watterson, K. 1994. *World Cup soccer gains momentum (Interview with Angelo Alogfu of World Cup USA)*, Data Based Advisor, 05/01/94, 120.

4.4 EMPIRICAL EVIDENCE OF TRUST MANAGEMENT

This section presents the results of the empirical study in this book. To investigate empirical evidence of trust management, a survey was conducted in 2006 with managerial personnel responsible for decision-making from IT departments in the top Polish (TOP-2004) companies from a broad range of industry sectors in:

- manufacturing
- trade and services
- finances.

The ranking of TOP-2004 was prepared for this occasion by a group of outstanding experts in economy within *Rzeczpospolita* magazine.

The main goal of the study was to verify empirically the existence and utilization of trust management tools in firms. A sample was chosen from the ranking "TOP 500"-2004 for the reason of strong competitive position on the market of firms listed there. Of the 500 firms receiving the online questionnaire from February to August of 2006, twenty-two (4.4%) responded. The results of the survey will be presented in figures and graphs.

First, the impact of trust on organizational success was investigated (see Graph 4.3, where the height of the "piggy bank" illustrates the frequency of a chosen answer concerning the influence of trust on organization success). Respondents were supposed to mark the magnitude of the impact of trust on the organization's success.

The findings confirmed a theoretical conclusion drawn from the discussion in Chapter 1, in the section "Theoretical Approaches toward Trust," referring to the paramount importance of trust in determining the prosperity of modern organizations. Trust seemed for 95% of respondents to influence organizational performance to a large extent. Almost 60% of respondents perceived trust as the factor contributing to positive outcomes of an organization to a very high degree. A further 35% of the responding firms acknowledged the strong support of trust in attainment of organizational goals. Undoubtedly, those findings provide empirical support for the theoretical conjecture concerning the strong relationship between trust and organizational success. Partners trusting each other are more likely to share ideas, discuss freely, keep promises, and meet deadlines. Those factors in turn play instrumental roles in shaping business relations. Relations based on trust reduce safeguards

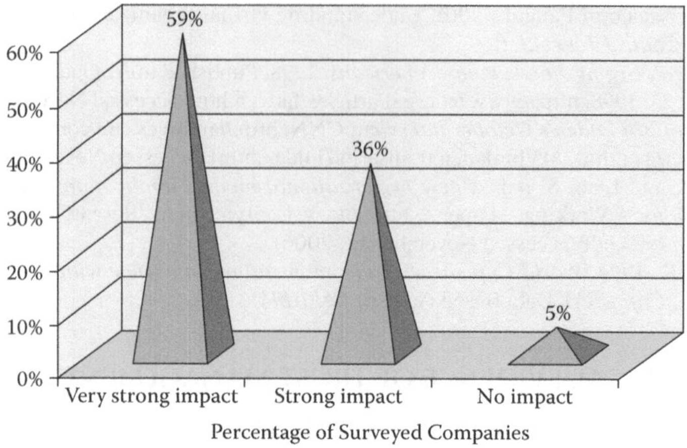

GRAPH 4.3 What impact does trust have on the organization's success?

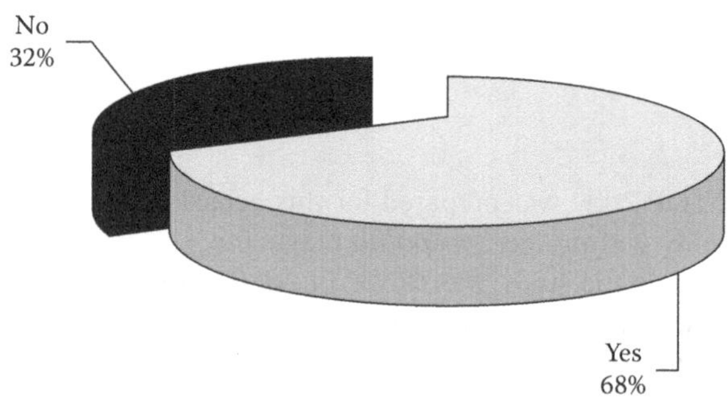

GRAPH 4.4 Does your company use any techniques/methods to evaluate business partners' trustworthiness?

(as shall be discussed later on) which, if all other things remain equal, leads to a decrease in transaction costs and consequently to an increase in profits.

A logical consequence of trust significance acknowledgment with regard to an organization's activities is the employment of particular techniques or methods within this scope (Graph 4.4). Almost 70% of firms reported the use of some techniques or methods to asses the trustworthiness of their business partners. In the remaining 32% of surveyed firms, this does not happen. The reason for this might lie in the simple fact that firms are not aware of the necessity of such an evaluation. Another explanation is that business relations might be rooted in good will and not in calculation in a strict sense. The assessment of trustworthiness is especially important whenever a decision is made to trust. However, in the case of a long-term and permanent cooperation, it might not play a prime role. This might explain why firms do not use formal techniques of assessing partners' trustworthiness.

Let's take a closer look at the results concerning trust placement among the responding firms. This process can take place with different frequencies (Graph 4.5).

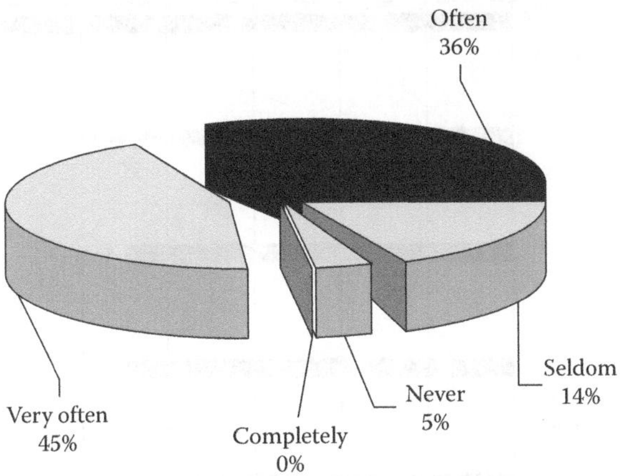

GRAPH 4.5 How often does your company trust its partners?

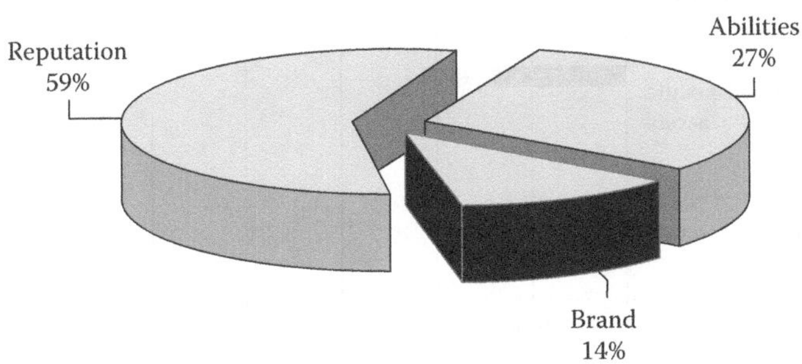

GRAPH 4.6 Which of the factors in this graph is the most important in the process of building trust between partners?

The majority of firms (81%) often or very often trusted their business partners. Only 14% seldom trusted their partners. The attitude of distrust toward business partners was a hallmark of 5% of surveyed firms. Moreover, respondents' answers shed light on the fact that there is no such thing as complete trust between business partners. This again is consistent with a theoretical presupposition.

In conclusion, we can say that trust is the foundation for business relations and is often likely to appear in cooperation. Trust can emerge from the positive appraisal of a partner's reputation, brand, and abilities. Familiarity with the significance of those elements might be valuable in the process of trust building (Graph 4.6).

In the view of almost 60% of respondents, reputation is the most important issue in the creation of trust between partners. Big companies build their reputations through a number of activities, such as brand creation, sponsoring, advertising, and public relations. Therefore, 14% of respondents believed that brand principally conceives trust. Reputation is a good tool for creating and evaluating the credibility of recent and potential business partners. Reputation can be verified; however, in assessing the trustworthiness

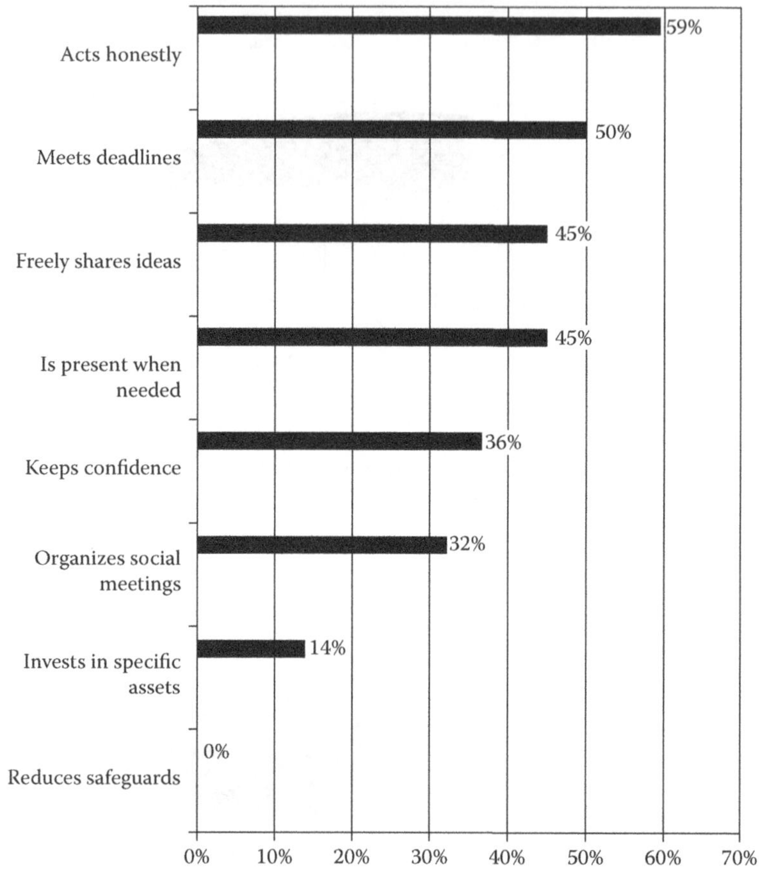

GRAPH 4.7 What does your firm do to encourage trust in a new transaction partner? (Note that all answers do not add up to 100%, because respondents were able to provide multiple answers.)

of small- and medium-sized companies, the criterion of abilities is more useful. Moreover, abilities constitute competence-based trust. Thus, the uppermost role of abilities in the process of trust building was confirmed by 27% of respondents.

Note that the trust-building source depends on the stage of cooperation. Under the circumstances, the appropriate selection of trust-building methods and means is essential and nearly always determines the level of trust in interactions between partners in future (Graph 4.7).

To gain a new partner, 59% of firms reported exerting the strategy of integrity, 50% meet deadlines, 45% freely express ideas, 45% are present when needed, 32% organize social meetings, and 14% invest in specific assets. An asset-specific investment engages the investor in the relationship and increases the exposition to opportunistic behavior. As the asset cannot be easily switched to another transaction, it is the discernible way to declare long-term commitment and dedication to a business partner as well as a technique for increasing predictability of interactions and

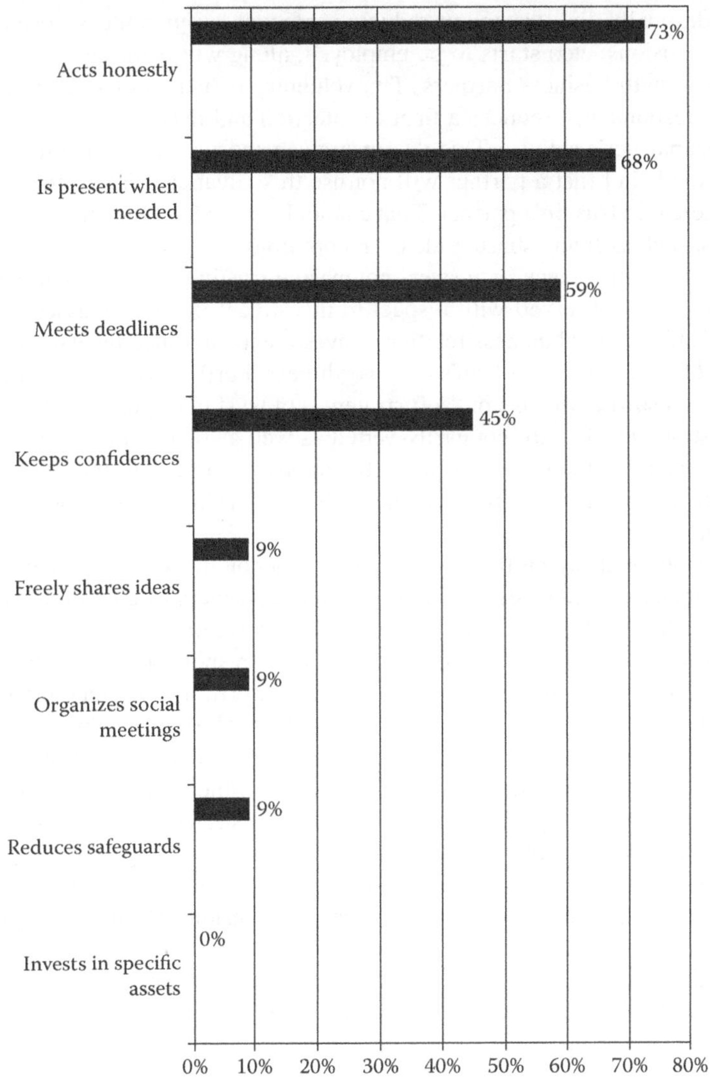

GRAPH 4.8 What does your firm do to enhance its trustworthiness in the eyes of its present business partners? (Note that all answers do not add up to 100%, because respondents were able to provide multiple answers.)

the sensation of safety. Yet, the strategy of reducing safeguards at the beginning is not applicable to new business partners.

Once a certain level of trust is established, it is still important to behave honestly (73%), be present when needed (68%), and meet deadlines (59%), as presented in Graph 4.8.

As shown in Graph 4.8, the items "freely share ideas" (9%) and "organize social meeting" (9%) with respect to cooperation were chosen less frequently for business partners already known from previous interactions than in the case of new partners. Additionally, the strategy of reducing safeguards (guarantees, formal control, lawsuit provisions, etc.) starts to be employed, along with gaining experience from cooperation with business partners. The voluntary reduction of safeguards (chosen by 9% of respondents) reduces a firm's protection and makes it more vulnerable to a business partner's actions. This demonstrates exposure to opportunistic behavior but with the belief that a partner will not use this advantage. Thus, this is a visible sign of increased trust in a partner. That cannot be said about the other trust-building strategies, such as freely sharing ideas and opinions.

The largest difference in answers comparing trusting relations with a new business partner was observed with respect to investment in specific assets. At a more advanced stage in the business relations, investments in a specific asset are not seen as a way of strengthening a firm's trustworthiness. Further continuation of cooperation entails a shift to the use of "soft strategies of trust management," which pertain to those strategies that are not easily vetted as well as being objectively asserted to exist (for example, the item "act honestly" in case of a new partner was chosen 13 times, whereas in case of a partner known from previous collaboration was chosen 16 times).

It goes without saying that trust is a phenomenon marked by subjectivism. The level of subjectivity depends on the perception of partners by a certain firm. So, the next issue addressed in the survey was a firm's perception of its partners in terms of their trustworthiness (Graph 4.9). The majority of respondents (73%) believed that their firm's business partners augment their trustworthiness by meeting deadlines; this response was followed by acting honestly (68%). The reason why acting honestly was not chosen by most of the respondents, as in the previous question, may reflect the simple fact that respondents find it hard to judge which of their business partners' activities might be recognized as honest. Thus, this strategy is not chosen most frequently by respondents. In contrast, easy to identify is "meeting deadlines," which is an apparent sign of consequence and reliability of a business partner.

Although it is difficult to arrive at general conclusions based on this pilot study, the results of the study of trust in efficient companies are promising. The overall results indicate empirical evidence on trust management application in those firms. Moreover, relationships based on mutual trust seem to contribute to their better performance. There are at least two reasons for this:

1. Encouragement of direct, honest interaction and commitment
2. Reduction of expensive safeguards.

Here, the authors would like to point out restrictions of using the results of the pilot study. These restrictions come from the sample size. The pilot study aimed to signal the ampleness and complexity of trust management. This is a first attempt to examine the phenomenon of trust in organizations from a management perspective. Additionally, the results of the pilot study confirmed the theoretical expectations described in the previous sections of this book.

To talk about trust building, we must remember the dynamic nature of trust and, therefore, take into consideration the present stage of the cooperation. For practitioners, there are a lot of issues to be pondered when formulating the strategy of

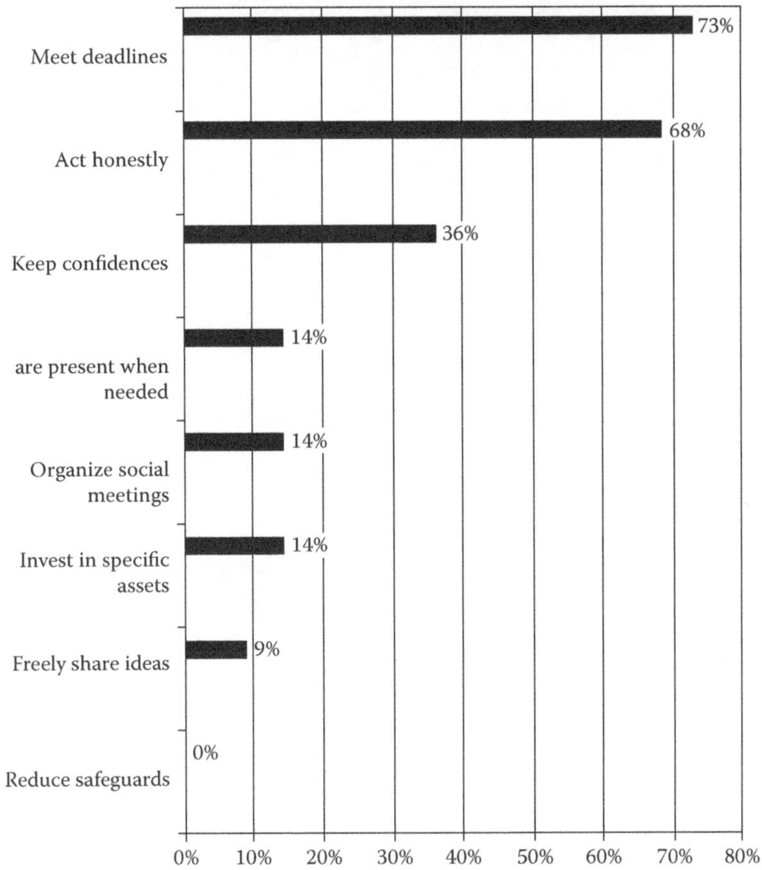

GRAPH 4.9 What do your firm's business partners do to enhance their trustworthiness? (Note that all answers do not add up to 100%, because respondents were able to provide multiple answers.)

business interactions (e.g., focus on clear communication, appropriate feedback, system of trust signals, and enhancement at each stage of cooperation).

From a practical point of view, further research in the field of trust management is justifiable, even desirable. Hence, this is a domain with a lack of systematization and empirical support of theoretical discussions, which limits its practical implications. Trust management is an interesting and thought-provoking field for further, more detailed research.

To recapitulate, the book on trust management in VOs was created by innovative people with a feeling for a future management tool—trust management—and hope to be an attraction for industrialists. For academics with an interest in the application of trust, we suggest further research to the benefits of societies, to their economies, and to industry in particular.

Index

F

G

H

M

N

O

Q

R

S

T

U

V

W

Z